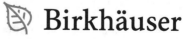

Advances in Mathematical Fluid Mechanics

Lecture Notes in Mathematical Fluid Mechanics

Editor-in-Chief
Giovanni P Galdi
University of Pittsburgh, Pittsburgh, PA, USA

Series Editors
Didier Bresch
Université Savoie-Mont Blanc, Le Bourget du Lac, France

Volker John
Weierstrass Institute, Berlin, Germany

Matthias Hieber
Technische Universität Darmstadt, Darmstadt, Germany

Igor Kukavica
University of Southern California, Los Angles, CA, USA

James Robinson
University of Warwick, Coventry, UK

Yoshihiro Shibata
Waseda University, Tokyo, Japan

Lecture Notes in Mathematical Fluid Mechanics as a subseries of "Advances in Mathematical Fluid Mechanics" is a forum for the publication of high quality monothematic work as well lectures on a new field or presentations of a new angle on the mathematical theory of fluid mechanics, with special regards to the Navier-Stokes equations and other significant viscous and inviscid fluid models.

In particular, mathematical aspects of computational methods and of applications to science and engineering are welcome as an important part of the theory as well as works in related areas of mathematics that have a direct bearing on fluid mechanics.

More information about this subseries at http://www.springer.com/series/15480

Joanna Rencławowicz
Wojciech M. Zajączkowski

The Large Flux Problem to the Navier-Stokes Equations

Global Strong Solutions in Cylindrical Domains

 Birkhäuser

Joanna Rencławowicz
Institute of Mathematics
Polish Academy of Sciences
Warsaw, Poland

Wojciech M. Zajączkowski
Institute of Mathematics
Polish Academy of Sciences
Warsaw, Poland

Institute of Mathematics
and Cryptology
Military University of Technology
Warsaw, Poland

ISSN 2297-0320   ISSN 2297-0339   (electronic)
Advances in Mathematical Fluid Mechanics
ISSN 2510-1374   ISSN 2510-1382   (electronic)
Lecture Notes in Mathematical Fluid Mechanics
ISBN 978-3-030-32329-5   ISBN 978-3-030-32330-1   (eBook)
https://doi.org/10.1007/978-3-030-32330-1

Mathematics Subject Classification: 35Q30, 76D03, 76D05, 35A01, 35B65, 35B45, 35D30, 35D35, 35G61

This book is published under the imprint Birkhäuser, www.birkhauser-science.com by the registered company Springer Nature Switzerland AG.
The registered company address is: Gewerbestrasse 11, 6330 Cham, Switzerland

Contents

Chapter 1
Introduction

Abstract This chapter is an introduction, where we describe the problem and define some key parameters to formulate main theorems: Theorems 1.1 and 1.2. We consider incompressible Navier-Stokes equations in cylindrical domain Ω with large inflow and outflow on the bottom and the top of the cylinder and the slip boundary conditions on the lateral part of the boundary. In order to prove the global existence of (v, p), where v is a velocity and p a pressure, with arbitrary large flux d, we show first the existence of weak solution, next we find conditions guaranteeing regularity of weak solution for large time T, and finally we achieve the existence of global regular solutions. In Theorem 1.1 we conclude that for sufficiently small parameter $\Lambda_2(T)$, which depends on data: norms of flux derivatives tangent to the bottom and the top of the cylinder, the initial condition $h(0) = v_{,x_3}(0)$, the derivative with respect to x_3 of force and with the estimate for weak solutions denoted with $\mathcal{A}$, we have $v, h \in W_2^{2,1}(\Omega^T), \nabla p, \nabla q = \nabla p_{,x_3} \in L_2(\Omega^T)$. In Theorem 1.2, for sufficiently large time T and from decay property of the equations we prove the existence of solutions in the interval $[kT, (k+1)T]$, $k \in \mathbb{N}_0$, therefore we can extend solutions step by step.

In this book, the motion of incompressible fluid described by the Navier-Stokes equations with large inflow and outflow is considered. The aim is to prove the existence of global regular solutions with arbitrary large flux. The domain is a straight non-axially symmetric cylinder with arbitrary cross-section. We assume the slip boundary conditions on the lateral part of the boundary, whereas on the bottom and the top of the cylinder there are some inflow and outflow fluxes. There is no restriction on the magnitude of the flux and we admit arbitrary large L_2 norm of initial velocity, but some homogeneity of the flux is necessary. Moreover, the initial velocity does not change too much along the axis of cylinder and the inflow does not change much along the directions perpendicular to this axis either with respect to time.

© Springer Nature Switzerland AG 2019

J. Renclawowicz, W. M. Zajączkowski, *The Large Flux Problem to the Navier-Stokes Equations*, Advances in Mathematical Fluid Mechanics, https://doi.org/10.1007/978-3-030-32330-1_1

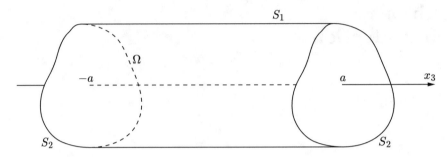

Fig. 1.1 Domain Ω

Physically, the problem can describe a motion of blood through straight parts of arteries. It seems that slip boundary conditions are more appropriate than nonslip because the blood can slip on the boundary. The model is the first step for further analysis of inflow-outflow problems and next, flows around some obstacles, with large velocities. This can also be the starting point to describe blood motion because walls of arteries are in fact elastic, so the problem with free boundary seems to be more natural: this will be the next step.

We consider the following initial boundary value problem to the Navier-Stokes equations in a cylindrical domain Ω with boundary $\partial\Omega = S = S_1 \cup S_2$ (Fig. 1.1) and with inflow and outflow

$$v_t + v \cdot \nabla v - \operatorname{div} \mathbb{T}(v,p) = f \qquad \text{in } \Omega^T = \Omega \times (0,T),$$

$$\operatorname{div} v = 0 \qquad \text{in } \Omega^T,$$

$$v \cdot \bar{n} = 0 \qquad \text{on } S_1^T = S_1 \times (0,T),$$

$$\nu\bar{n} \cdot \mathbb{D}(v) \cdot \bar{\tau}_\alpha + \gamma v \cdot \bar{\tau}_\alpha = 0, \quad \alpha = 1,2 \qquad \text{on } S_1^T, \quad (1.1)$$

$$v \cdot \bar{n} = d \qquad \text{on } S_2^T = S_2 \times (0,T),$$

$$\bar{n} \cdot \mathbb{D}(v) \cdot \bar{\tau}_\alpha = 0, \quad \alpha = 1,2 \qquad \text{on } S_2^T,$$

$$v|_{t=0} = v(0) \qquad \text{in } \Omega,$$

where the stress tensor has the form

$$\mathbb{T}(v,p) = \nu\mathbb{D}(v) - p\mathbb{I}.$$

By $\nu > 0$ we denote the constant viscosity coefficient, $\gamma > 0$ is the slip coefficient, $\bar{n}$ is the unit outward vector normal to S, τ_α, $\alpha = 1,2$, are vectors tangent to S, $\mathbb{I}$ is the unit matrix, and $\mathbb{D}(v)$ is the dilatation tensor of the form

$$\mathbb{D}(v) = \{v_{i,x_j} + v_{j,x_i}\}_{i,j=1,2,3}.$$

We set $S = S_1 \cup S_2$, where S_1 is parallel to the x_3 axis and S_2 is perpendicular to it. Hence

$$
\begin{aligned}
S_1 &= \{x \in \mathbb{R}^3 : \varphi_0(x_1, x_2) = c_0, \ -a < x_3 < a\}, \\
S_2(-a) &= \{x \in \mathbb{R}^3 : \varphi_0(x_1, x_2) < c_0, \ x_3 = -a\}, \\
S_2(a) &= \{x \in \mathbb{R}^3 : \varphi_0(x_1, x_2) < c_0, \ x_3 = a\},
\end{aligned}
\tag{1.2}
$$

where a, c_0 are given positive numbers and $\varphi_0(x_1, x_2) = c_0$ describes a sufficiently smooth closed curve in the plane $x_3 = \text{const}$.

To describe the inflow and outflow we define

$$
\begin{aligned}
d_1 &= -v \cdot \bar{n}|_{S_2(-a)}, \\
d_2 &= v \cdot \bar{n}|_{S_2(a)},
\end{aligned}
\tag{1.3}
$$

with $d_i \geq 0$, $i = 1, 2$. Equation $(1.1)_2$ implies the following compatibility condition

$$\int_{S_2(-a)} d_1 dS_2 = \int_{S_2(a)} d_2 dS_2. \tag{1.4}$$

The goal of this book is to prove the existence of global regular solutions to problem (1.1)–(1.4) with arbitrary large flux d. In order to demonstrate such results we are going to proceed in three main steps: first, we show the existence of weak solution, next we find the conditions guaranteeing regularity of weak solution for large time, and finally we achieve the existence of global regular solutions.

This book consists of 12 chapters. In Chap. 2 notation and auxiliary results are introduced.

In Chap. 3 the energy type estimate for weak solutions to problem (1.1)–(1.4) is derived. The main result is formulated in Proposition 3.4. Since (1.1)–(1.4) is an inflow-outflow problem the normal component of velocity on S_2 does not vanish (see $(1.1)_5$). This makes the direct derivation of energy type estimate for solutions to (1.1)–(1.4) impossible. To make it possible we introduce the Hopf function (3.2) which help us to transform the problem (1.1)–(1.4) into the new one, with homogeneous Dirichlet boundary conditions. Namely, we construct some extension δ such that $w = v - \delta$ (see (3.7)) satisfies that $w \cdot \bar{n}|_S = 0$. Since w is a solution to problem (3.8) the energy type inequality (3.10) can be derived by integration by parts. Therefore, in Lemma 3.2, we are able to obtain the energy type estimate for solutions of problem (3.8). However, to obtain an energy estimate we need some estimates in weighted Sobolev spaces for the extension δ

(see (3.12), (3.13), and (3.14)). In fact we need estimates in these weighted spaces for solutions to the auxiliary Neumann problem for the Poisson equation (3.6). The existence of solutions to (3.6) in L_2-weighted spaces is shown in Chap. 10 and in L_p-weighted spaces in Chap. 11. The weight is equal to a power function of the distance to S_2 which is not a Muckenhoupt weight.

In reality, in Proposition 3.4 we prove global estimate using a step by step in time approach. It is shown that $\|v\|_{V(kT,t)}$, $t \in [kT, (k+1)T]$, $k \in \mathbb{N}_0$, where the norm $V(kT,t)$ is defined in (2.3), is bounded by a constant depending on data but independent of k. We have to emphasize that this estimate is proved in detail and that we derive a priori estimates only, while the existence is shown briefly. The existence of weak solutions is then proved by the Faedo-Galerkin method in Chap. 12, Theorems 12.1 and 12.2 and in detail in Renclawowicz and Zajączkowski [RZ3] (see also arguments by the Leray-Schauder fixed point theorem in [RZ4, Z5]). The step by step in time approach implies less restriction on data because only finite time integrals appear in the bound.

In Chap. 3 there is also derived an a priori bound for $\|v_t\|_{V(kT,t)}$, $t \in [kT, (k+1)T]$. The bound depends on $\|v\|_{2,2,\Omega \times (kT,(k+1)T)}$ (with the notation for Sobolev spaces and norms introduced in (2.2)) which is estimated in Chap. 4.

In Chap. 4 we find an estimate guaranteeing regularity of weak solutions for large finite time interval. In order to get this, we transform problem (1.1) into a series of problems. Then assuming smallness of the following data: $v_{,x_3}(0), f_{,x_3}(0), d_{,x'}$, where $x = (x_1, x_2, x_3)$ are the Cartesian coordinates such that x_3 is along the cylinder and $x' = (x_1, x_2)$, such estimate can be shown. Since the idea of the proof is little complicated we sketch its steps. We introduce quantities $h = v_{,x_3}$, $q = p_{,x_3}$, $g = f_{,x_3}$ which satisfy problem (4.6). In virtue of Lemmas 4.3, 4.4, and Corollary 4.5 we achieve the estimate

$$\|h\|_{V(\Omega^t)} \leq \varphi(D_1(t), \mathcal{V}(t))(\Lambda_1(t) + |h(0)|_{2,\Omega}), \tag{1.5}$$

where φ is an increasing positive function, and norms notation is given in (2.1), (2.2), and (2.3),

$$D_1(t) = |d_1|_{3,6,S_2^t}, \quad \mathcal{V}(t) = |\nabla v|_{3,2,\Omega^t} \qquad \text{(see (4.40)),}$$

$$\Lambda_1^2(T) = \int_0^T (\|d_{x'}\|_{1,S_2}^2 + \|d_t\|_{1,S_2}^2 + |f_3|_{4/3,S_2}^2 + |g|_{6/5,\Omega}^2)dt$$

$$+ \mathcal{A}^2 \sup_t \|d_{x'}\|_{1,3/2,S_2}^2 \qquad \text{(see (4.38)),} \tag{1.6}$$

$$\|v\|_{V(\Omega^t)}^2 \leq \varphi(\sup_t \|d\|_{1,3,S_2}) \int_0^t (|f|_{6/5,\Omega}^2 + \|d\|_{1,3,S_2}^2 + \|d_t\|_{1,6/5,S_2}^2)dt'$$

$$+ |v(0)|_{2,\Omega}^2 \equiv \mathcal{A}^2(t) \qquad \text{(see (4.13)).}$$

Next, in order to increase regularity of the first two components of velocity, i.e., $v' = (v_1, v_2)$, we introduce the vorticity $\chi(x,t) = v_{2,x_1} - v_{1,x_2}$ satisfying problem (4.9) (see Lemma 4.2). We are allowed to use vorticity since Ω is a cylindrical domain parallel to the x_3-axis. In Corollary 4.8 we obtain the inequality (4.54)

$$\|\chi\|^2_{V(\Omega^t)} \le c(1 + \mathcal{A}^2)(|h_3|^2_{3,\infty,\Omega^t} + |F_3|^2_{6/5,2,\Omega^t}$$
$$+ \|v'\|^2_{W_2^{1,1/2}(\Omega^t)} + \|v'\|^2_{5/6,2,\infty,\Omega^t} + |\chi(0)|^2_{2,\Omega}).$$

The above inequality can be attained because slip boundary conditions are used.

Then, from Lemmas 4.9 and 4.10 and sufficiently small $\Lambda_2 = \Lambda_1 + |h(0)|_{2,\Omega}$ (see (4.61)) we derive the inequality

$$\|v\|_{W_{5/3}^{2,1}(\Omega^T)} \le \varphi(\mathcal{A})H_2 + \varphi(D),$$

where D describes all data norms, $\mathcal{A} = \mathcal{A}(t)$ is the estimate for weak solution given in (4.13), and

$$H_2 = H_1 + |h|_{10/3,\Omega^t},$$
$$H_1 = \sup_t |h|_{3,\Omega} + \|h\|_{1,2,\Omega^t}.$$

Using the estimate for the Stokes system (4.63), Lemma 4.12 implies

$$\|v\|_{W_2^{2,1}(\Omega^t)} + |\nabla p|_{2,\Omega^t} \le \varphi(D, H_2). \tag{1.7}$$

Applying the Stokes system connected with problem (4.6) and smallness of $\Lambda_2(T) = \Lambda_1(T) + |h(0)|_{2,\Omega}$, Lemma 4.13 implies the estimate

$$\|h\|_{W_\sigma^{2,1}(\Omega^t)} + \|\nabla q\|_{L_\sigma(\Omega^t)} \le \varphi(D), \quad \frac{5}{3} \le \sigma \le \frac{10}{3}. \tag{1.8}$$

Moreover, the imbedding

$$H_2 \le c\|h\|_{W_\sigma^{2,1}(\Omega^t)}, \text{ for } \sigma \ge \frac{5}{3}$$

holds. Therefore, combining this with (1.7) and (1.8) we conclude

$$\|v\|_{W_2^{2,1}(\Omega^t)} + |\nabla p|_{2,\Omega^t} \le \varphi(D). \tag{1.9}$$

The above considerations in Chap. 4 imply bounds for the quantities

$$\|v\|_{W_2^{2,1}(\Omega^T)}, \quad \|h\|_{W_2^{2,1}(\Omega^T)},$$

for sufficiently small $\Lambda_2(T)$, where T can be chosen as large as we want (see definition of Λ_1 in (1.6)). This means that in Chap. 4 long-time estimates are proved. Applying the Leray-Schauder fixed point theorem we prove long-time existence of regular solutions to (1.1)–(1.4) and (4.6). For application of the Leray-Schauder fixed theorem see Chap. 12 and [RZ4], [Z5].

Estimates (1.8) and (1.9) hold for any $t \in \mathbb{R}_+$. On the other hand, D depends on time integrals of some quantities, so for large t they must decay sufficiently fast. Considering estimates (1.8) and (1.9) in time interval $(0, T)$ we show, in Chaps. 5–8, the existence of T sufficiently large so that (1.13) hold, i.e.

$$\|v((k+1)T)\|_{1,\Omega} \le \|v(kT)\|_{1,\Omega},$$

$$\|h((k+1)T)\|_{1,\Omega} \le \|h(kT)\|_{1,\Omega}, \quad \text{for any } k \in \mathbb{N}_0.$$

These inequalities imply the existence of global regular solutions to problem (1.1) for $t \in \mathbb{R}_+$. Hence, we have the following.

Theorem 1.1 *Consider problems (1.1)–(1.4) for v and (4.6) for $h = v_{,x_3}$, where the domain Ω is non-axially symmetric. Let time $T > 0$ be given and denote $q = p_{,x_3}$, $F_3 = f_{2,x_1} - f_{1,x_2}$, and $g = f_{,x_3}$. Assume that $\Lambda_2(T) = \Lambda_1(T) + |h(0)|_{2,\Omega}$, where $\Lambda_1(T)$ is given by (1.6), is sufficiently small. Additionally, we require that:*

- *external forces satisfy $f, g \in L_2(\Omega^T)$ and $F_3 \in L_{6/5,2}(\Omega^T)$,*
- *initial velocity satisfies $v(0), h(0) \in H^1(\Omega)$,*
- *flux satisfies $d \in L_\infty(0, T; W_3^1(S_2)) \cap W_2^{3/2,3/4}(S_2^T)$, $d_{x'} \in W_2^{3/2,3/4}(S_2^T)$, $d_t \in L_2(0, T; W_{6/5}^1(S_2))$.*

Then there exists a solution such that $v, h \in W_2^{2,1}(\Omega^T)$, $\nabla p, \nabla q \in L_2(\Omega^T)$ and the following estimate holds

$$\|v\|_{W_2^{2,1}(\Omega^T)} + \|h\|_{W_2^{2,1}(\Omega^T)} + |\nabla p|_{2,\Omega^T} + |\nabla q|_{2,\Omega^T}$$

$$\le \varphi\bigg(|f|_{2,\Omega^T}, |g|_{2,\Omega^T}, |F_3|_{6/5,2,\Omega^T}, \|v(0)\|_{1,\Omega}, \|h(0)\|_{1,\Omega}, \tag{1.10}$$

$$\sup_{t \le T} \|d\|_{1,3,S_2}, \|d\|_{W_2^{3/2,3/4}(S_2^T)}, \|d_t\|_{L_2(0,T;W_{6/5}^1(S_2))}, \|d_{x'}\|_{W_2^{3/2,3/4}(S_2^T)}\bigg),$$

where φ is an increasing positive function and $x' = (x_1, x_2)$.

In Chap. 5 we prove the global existence of solutions to problem (1.1)–(1.4), (4.6) by applying the step by step in time argument.

From Lemma 5.2 we have (see (5.13))

$$\|v(kT)\|_{1,\Omega} \le Q_1(T) + \exp(-\nu kT)\|v(0)\|_{1,\Omega} \tag{1.11}$$

and Lemma 5.4 implies (see (5.14))

$$\|h(kT)\|_{1,\Omega} \le Q_2(T) + \exp(-\nu kT)\|h(0)\|_{1,\Omega}, \tag{1.12}$$

where $Q_1(T), Q_2(T)$ are equal to the first terms on the r.h.s. of (5.13) and (5.14), respectively. Inequalities (1.11) and (1.12) follow from the considerations from Chaps. 6 to 9.

From (1.11), (1.12) for T sufficiently large and some $\|v(0)\|_{1,\Omega}, \|h(0)\|_{1,\Omega}$ we obtain

$$\|v((k+1)T)\|_{1,\Omega} \le \|v(kT)\|_{1,\Omega},$$
$$\|h((k+1)T)\|_{1,\Omega} \le \|h(kT)\|_{1,\Omega}. \tag{1.13}$$

Therefore we have the following.

Theorem 1.2 *Let the assumptions of Theorem 1.1 hold. Then for T sufficiently large and conditions implying (1.13) we have the existence of solutions to problems (1.1)–(1.4) and (4.6) in the interval $[kT, (k+1)T]$, $k \in \mathbb{N}_0$ such that, with $\Omega \times (kT, (k+1)T) = \Omega^{kT}$, $S_2 \times (kT, (k+1)T) = S_2^{kT}$,*

$$\|v\|_{W_2^{2,1}(\Omega^{kT})} + \|h\|_{W_2^{2,1}(\Omega^{kT})} + |\nabla p|_{2,\Omega^{kT}} + |\nabla q|_{2,\Omega^{kT}} \le \varphi(D(kT, (k+1)T)),$$

where

$$D(kT, (k+1)T) \equiv \Big[|f|_{2,\Omega^{kT}}, |g|_{2,\Omega^{kT}}, |F_3|_{6/5,2,\Omega^{kT}}, \|v(0)\|_{1,\Omega}, \|h(0)\|_{1,\Omega},$$

$$\sup_{kT \le t \le (k+1)T} \|d(t)\|_{1,3,S_2}, \|d\|_{W_2^{3/2,3/4}(S_2^{kT})},$$

$$\|d_t\|_{L_2(kT,(k+1)T;W_{6/5}^1(S_2))}, \|d_{x'}\|_{W_2^{3/2,3/4}(S_2^{kT})} \Big].$$

The aim of this monograph is to prove the existence of global regular solutions to the Navier-Stokes equations in a bounded straight cylinder with large inflow and outflow. Our approach is strictly connected with the structure of the Navier-Stokes system and is motivated by the classical paper of Ladyzhenskaya [L1]. Using ideas of [L1], W.M. Zajączkowski came to the conclusion that stability of $2d$ solutions was possible. Hence the global regular solutions close to $2d$ solutions are proved in [Z5, Z6, Z7, Z8]. To show the existence of such solutions we need geometrical restrictions such that the considered domain must be a straight cylinder. We need also the slip boundary conditions on the part of the boundary because with such conditions a problem for $2d$ rotation can be applied (see also [Z5, Z6]).

The case with inflow-outflow was considered in Rencławowicz and Zajączkowski papers [RZ3, RZ4, RZ5, RZ6] and in case of Poiseuille flow in [PZ]. In [RZ3] the existence of weak solutions to problem (1.1) is proved. The proof bases on the following steps.

1. Appropriate extension of the nonhomogeneous boundary conditions on S_2.
2. A global estimate for solutions to (1.1) is proved by using some weighted estimates.
3. Once we have the estimate, the existence of weak solutions is proved by the Faedo-Galerkin method.

Under the smallness conditions presented in Theorem 1.1, the long-time existence of solutions to problem (1.1) was proved in [RZ4]. In [RZ5, RZ6] the global existence of solutions to problem (1.1) formulated in Theorem 1.2 was proved. However, we can clarify some aspects of the proof presented there.

The global existence of solutions to problem (1.1) without inflow and outflow was shown in [Z5, RZ7] under smallness of the following quantity

$$\Lambda(T) = |f_3|^2_{4/3,S_2^T} + |g|^2_{6/5,\Omega^T} + |h(0)|^2_{2,\Omega}.$$

We mention also papers [SZ1, SZ2, SZ3] where problem (1.1) without inflow and outflow but coupled with the heat convection is considered.

There is a huge literature concerning on the existence and regularity of solutions to the Navier-Stokes equations, see [S, Le, G2].

In [S], Sohr developed a very advanced and beautiful theory of semigroup with applications to the Navier-Stokes equations. Unfortunately, such theory does not feel the structure of the Navier-Stokes system, so in reality it can be applied to prove only either local existence or global existence with small data.

In [Le], Lemarié-Rieusset presented a lot of spaces (Besov, Lorentz, Morrey and Sobolev) and a lot of different techniques and approaches applied to the Navier-Stokes equations. He also mentioned the idea of semigroup. Nevertheless, the chapter corresponding to the global existence of regular solutions is very short (see Ch. 33). The book is indeed a vast and deep review of different techniques, spaces, and methods which can be used in the problems of Navier-Stokes equations but it is not adequate in our case.

The new book of Lemarie-Rieusset, [Le2], is some extension of the previous one. The author recalls properties of mild solutions to Navier-Stokes equations in $H^s, s \geq 1/2$, $L_p, p \geq 3$ and Besov, Morrey spaces. Moreover, the author describes the regularity criterion of Caffarelli-Kohn-Nirenberg and L_3-criterion of Escaurioza-Seregin-Sverak on suitable weak solutions. However, this book does not contain any results on the existence of global regular solutions close to either axisymmetric or two-dimensional solutions. The new book of Robinson et al. [RRS] is focused on the regularity problem for solutions to the Navier-Stokes equations. The authors distinguish the following topics: the existence of global-in-time weak solutions of Leray-Hopf problem, the local regularity theory originated from the Serrin criterion, and partial regularity developed by Scheffer, Caffarelli, Kohn, and Nirenberg. The book does not contain any results on the existence of global special regular solutions (to compare, see [Z6, Z11, Z12]).

Since the existence of regular weak solutions to the Navier-Stokes equations is still an open problem we can distinguish three kinds of possibilities of showing the existence of global regular non-small solutions.

1. The existence of global regular solutions close to the two-dimensional solutions proved by Ladyzhenskaya in [L2]. Such solutions are shown in [RZ7], [Z5].
2. In [Z11, Z12] the existence of global regular solutions is established, which are close to the axially symmetric solutions with vanishing angular coordinate of velocity. The existence of corresponding axially symmetric solutions was proved by Ladyzhenskaya [L3] (see also [UY]).
3. The existence of regular global solutions to the Navier-Stokes equations describing fast rotating fluids (see papers of Babin et al.: [BMN1, BMN2, BMN3], Mahalov and Nicolaenko [MN], and Farwig et al.: [FST]).

To prove the existence of global regular solutions we need a method that is not sensitive to nonlinearity, and therefore the most appropriate approach is the energy method and L_2-type Sobolev spaces. Although it seems old-fashioned, it is the proper one.

Motivated by the results of Chap. 8 from the book of Galdi [G2], where the inflow-outflow problem for stationary Navier-Stokes is considered, we started to consider the nonstationary problem in the cylinder. We have to mention that Takeshita counter example, see [T], does not work in our case.

Chapter 2
Notation and Auxiliary Results

Abstract In this chapter we introduce the basic notation for classical norms (Lebesque spaces L_p and Sobolev spaces W_p^s and H^s) and define some more complicated spaces and norms:

- anisotropic Sobolev spaces on Ω and Q^T for $\Omega, Q \subset \mathbb{R}^3$: $W_{p_1,p_2}^k(\Omega)$, $W_r^{l,l/2}(Q^T)$,
- anisotropic Sobolev spaces with mixed norms: $W_{p_1,p_2}^{l,l/2}(\Omega^T)$,
- energy type spaces: $V^s(\Omega^T), V(t_1, t_2)$,
- Sobolev-Slobodetskii spaces: $W_r^{l,l/2}(S^T)$,
- Besov and Nikolskii spaces: $B_{r,\theta}^{l,l_0}(G), H_r^{l,l_0}(G)$,
- weighted spaces: $L_{p,\mu}^k(\Omega), H_\mu^k(\Omega_\varrho), V_{p,\beta}^l(Q)$.

We consider some auxiliary problems in cylindrical domains (Poisson equation and stationary Stokes system) and calculate corresponding Green functions. We also construct partition of unity and collect some facts useful in the next chapters: trace and extensions theorems, imbeddings and interpolation inequalities, and Korn inequality.

2.1 Spaces and Basic Notation

First we introduce the simplified notation for standard Lebesgue spaces L_p and Sobolev spaces W_p^s.

Definition 2.1 (Lebesque and Sobolev spaces) (see [AF]) We set the following notation for Lebesque spaces

$$\|u\|_{L_p(Q)} = |u|_{p,Q}, \quad \|u\|_{L_p(Q^t)} = |u|_{p,Q^t},$$

$$\|u\|_{L_q(0,t;L_p(Q))} = |u|_{p,q,Q^t}, \tag{2.1}$$

© Springer Nature Switzerland AG 2019
J. Renclawowicz, W. M. Zajączkowski, *The Large Flux Problem to the Navier-Stokes Equations*, Advances in Mathematical Fluid Mechanics, https://doi.org/10.1007/978-3-030-32330-1_2

and Sobolev spaces

$$\|u\|_{H^s(Q)} = \|u\|_{s,Q}, \quad \text{where } H^s(Q) = W_2^s(Q),$$
$$\|u\|_{W_p^s(Q)} = \|u\|_{s,p,Q}, \tag{2.2}$$

where Q is a bounded set in $\mathbb{R}^n$ (in particular, $Q = \Omega$ or $Q = S \equiv \partial\Omega$) or $Q = \mathbb{R}^n$, $Q^t = Q \times (0, t)$, $p, q \in [1, \infty)$, $s \in \mathbb{R}$.

In definitions below we use the notation: $D_x^\alpha = \partial_{x_1}^{\alpha_1} \partial_{x_2}^{\alpha_2} \partial_{x_3}^{\alpha_3}$, $D_{x'}^{\alpha'} = \partial_{x_1}^{\alpha_1} \partial_{x_2}^{\alpha_2}$, where $\alpha' = (\alpha_1, \alpha_2)$ and $\alpha = (\alpha_1, \alpha_2, \alpha_3)$ are multiindices such that $\alpha_i \in \mathbb{N}_0$ and $|\alpha'| = \alpha_1 + \alpha_2$, $|\alpha| = \alpha_1 + \alpha_2 + \alpha_3$.

We also set

$$\|u\|_{L_q(0,t;W_p^k(\Omega))} = \|u\|_{k,p,q,\Omega^t},$$
$$\|u\|_{L_p(0,t;W_p^k(\Omega))} = \|u\|_{k,p,\Omega^t}.$$

Definition 2.2 (Weighted spaces) Let $\eta(x_3) = \min_{i=1,2} \text{dist}(x_3, S_2(a_i))$. Then we introduce weighted spaces by

$$\|u\|_{L_{p,\mu}^k(\Omega)} = \left(\sum_{|\alpha|=k} \int_\Omega |D_x^\alpha u|^p \eta^{p\mu} dx \right)^{1/p} < \infty, \quad \mu \in \mathbb{R}, \quad p \in (1, \infty),$$

and $L_{p,\mu}(\Omega) = L_{p,\mu}^0(\Omega)$.

In Chap. 10, in Definition 10.1 we also define $H_\mu^k(\Omega_\varrho)$ for $\Omega_\varrho = \{x \in \Omega : 0 < x_3 < \varrho\}$.

In Chap. 11, we also use more specific weighted spaces: $V_{p,\beta}^l(Q)$ for $Q \subset \mathbb{R}_+^3$ is a set of functions with the finite norm

$$\|u\|_{V_{p,\beta}^l(Q)} = \left(\sum_{|\alpha|\le l} \int_Q |D_x^\alpha u|^p x_3^{p(\beta+|\alpha|-l)} dx' dx_3 \right)^{1/p}, \quad p \in [1, \infty),$$

where $x' = (x_1, x_2)$, $\beta \in \mathbb{R}$, $\alpha = (\alpha_1, \alpha_2, \alpha_3)$ is a multiindex. Moreover,

$$V_{p,\beta}^0(Q) = L_{p,\beta}(Q), \quad V_{2,\beta}^l(Q) = H_\beta^l(Q),$$
$$\|u\|_{V_{\infty,\beta}^l(Q)} = \sum_{|\alpha|\le l} \text{ess} \sup_{x\in Q} \left(|D_x^\alpha u|^p x_3^{p(\beta+|\alpha|-l)} \right).$$

Definition 2.3 (Anisotropic Sobolev spaces on Ω) Introduce the anisotropic Sobolev space

$$\|u\|_{W_{p_1,p_2}^k(\Omega)} = \left[\int_{-a}^a \left(\int_{S_2} \sum_{|\alpha'|\le k} |D_{x'}^{\alpha'} u|^{p_1} dx_1 dx_2 \right)^{p_2/p_1} dx_3 \right]^{1/p_2} < \infty,$$

where $p_1, p_2 \in [1, \infty]$, $k \in \mathbb{N}_0 = \mathbb{N} \cup \{0\}$, $x' = (x_1, x_2)$. For $k = 0$ we have

$$\|u\|_{W^0_{p_1,p_2}(\Omega)} = \|u\|_{L_{p_1,p_2}(\Omega)} = |u|_{p_1,p_2,\Omega}.$$

For $p_1 = \infty$ or $p_2 = \infty$ we have, respectively

$$\|u\|_{W^k_{p_1,\infty}(\Omega)} = \operatorname*{ess\ sup}_{x_3 \in (-a,a)} \left(\int_{S_2} \sum_{|\alpha'| \le k} |D^{\alpha'}_{x'} u|^{p_1} dx_1 dx_2 \right)^{1/p_1} < \infty,$$

$$\|u\|_{W^k_{\infty,p_2}(\Omega)} = \operatorname*{ess\ sup}_{x' \in S_2} \left(\int_{-a}^{a} \sum_{|\alpha'| \le k} |D^{\alpha'}_{x'} u|^{p_2} dx_3 \right)^{1/p_2} < \infty,$$

$$\|u\|_{W^k_{\infty,\infty}(\Omega)} = \operatorname*{ess\ sup}_{x \in \Omega} \sum_{|\alpha'| \le k} |D^{\alpha'}_{x'} u| < \infty.$$

Definition 2.4 (Anisotropic Sobolev spaces on Q^T) By $W^{l,l/2}_r(Q^T)$, $Q \in \mathbb{R}^3$, where $r \in [1, \infty]$ and l is even, we denote the anisotropic Sobolev space with the following finite norm

$$\|u\|_{W^{l,l/2}_r(Q^T)} = \left(\sum_{\alpha + 2\alpha_0 \le l} \int_0^t \int_Q |D^\alpha_x \partial^{\alpha_0}_t u|^r dx dt \right)^{1/r}, \quad r \in [1, \infty)$$

$$\|u\|_{W^{l,l/2}_\infty(Q^T)} = \operatorname*{ess\ sup}_{(x,t) \in Q^T} \sum_{\alpha + 2\alpha_0 \le l} |D^\alpha_x \partial^{\alpha_0}_t u|.$$

We will apply also the following notation

$$\|u\|_{W^{2,1}_p(\Omega^t)} = \left(\int_{\Omega^t} (|\nabla^2 u|^p + |\partial_t u|^p + |u|^p) dx dt' \right)^{1/p} = \|u\|_{(2),p,\Omega^t}.$$

Definition 2.5 (Anisotropic Sobolev spaces with mixed norm) Introduce the anisotropic Sobolev space with the mixed norm: $W^{l,l/2}_{p_1,p_2}(\Omega^T)$, l-even, $p_1, p_2 \in [1, \infty]$, where

$$\|u\|_{W^{l,l/2}_{p_1,p_2}(\Omega^T)} = \sum_{|\alpha| \le l} \|D^\alpha_x u\|_{L_{p_2}(0,T;L_{p_1}(\Omega))} + \sum_{\alpha_0 \le l/2} \|\partial^{\alpha_0}_t u\|_{L_{p_2}(0,T;L_{p_1}(\Omega))}$$

$$+ \|u\|_{L_{p_2}(0,T;L_{p_1}(\Omega))},$$

where $\alpha = (\alpha_1, \alpha_2, \alpha_3)$, $|\alpha| = \alpha_1 + \alpha_2 + \alpha_3$, $\alpha_i \in \mathbb{N}_0$, $i = 1, 2, 3$.

To describe the energy type estimates for solutions to the Navier-Stokes equations we need the following spaces characteristic for parabolic equations.

Definition 2.6 (Energy Type spaces)

$$V^s(\Omega^T) = \{u: \text{ ess} \sup_{t \le T} \|u(t)\|_{s,\Omega} + \|\nabla u\|_{s,2,\Omega^T} < \infty\}, \quad s \in \mathbb{N}_0.$$

For $s = 0$, we have $V(\Omega^T) = V^0(\Omega^T)$ and

$$\|u\|_{V(\Omega^T)} = \text{ess} \sup_{t \le T} |u(t)|_{2,\Omega} + |\nabla u|_{2,\Omega^T}.$$

In the step by step in time approach we need spaces $V(\Omega \times (t_1, t_2)) = V(t_1, t_2)$, $t_1 < t_2$, with the finite form

$$\|u\|_{V(t_1,t_2)} = \text{ess} \sup_{t_1 \le t \le t_2} |u(t)|_{2,\Omega} + \left(\int_{t_1}^{t_2} |\nabla u(t')|_{2,\Omega}^2 dt'\right)^{1/2}. \quad (2.3)$$

In particular, for $t_1 = 0, t_2 = t$, we denote sometimes

$$V(0,t) = V(\Omega \times (0,t)) = V(\Omega^t).$$

To describe traces of u on the boundary S of Ω we introduce the Sobolev-Slobodetskii spaces $W_r^{l,l/2}(S^T)$, where l is non-integer. The norm of this space is the following (see Ladyzhenskaya et al. [LSU, Ch. 2, Sect. 3], Il'in and Solonnikov [IS], and Solonnikov [S2]).

Definition 2.7 (Sobolev-Slobodetskii spaces) For $r \in [1, \infty)$ and non-integer l we define space $W_r^{l,l/2}(S^T)$ with the finite norm

$$\|u\|_{W_r^{l,l/2}(S^T)} = \sum_{\alpha + 2\alpha_0 \le [l]} \|D_x^\alpha \partial_t^{\alpha_0} u\|_{L_r(S^T)}$$

$$+ \left(\sum_{\alpha + 2\alpha_0 \le [l]} \int_0^T \int_S \int_S \frac{|D_{x'}^\alpha \partial_t^{\alpha_0} u(x',t) - D_{x''}^\alpha \partial_t^{\alpha_0} u(x'',t)|^r}{|x' - x''|^{2+r(l-[l])}} dx' dx'' dt\right)^{1/r}$$

$$+ \left(\sum_{0 < l - \alpha - 2\alpha_0 \le 2} \int_0^T \int_0^T \int_S \frac{|D_x^\alpha \partial_{t'}^{\alpha_0} u(x,t') - D_x^\alpha \partial_{t''}^{\alpha_0} u(x,t'')|^r}{|t' - t''|^{1+\frac{r}{2}(l-[l])}} dt' dt'' dx\right)^{1/r}.$$

We are going to apply some theorems on imbedding and traces for the Sobolev-Slobodetskii spaces. It is convenient to use such results in the language of Besov spaces. Therefore we define them (see Besov et al. [BIN, Ch. 4, Sect. 18]).

Introduce the differences

$$\Delta_i(h)f(x) = f(x + e_i h) - f(x), \quad x \in \mathbb{R}^N, \ i = 1,\ldots,n$$

where e_i is the unit vector directed along the i-th Cartesian coordinate. Next

$$\Delta_i^k(h)f(x) = \Delta_i(\Delta_i^{k-1}(h)f(x)), \quad k \in \mathbb{N}.$$

For a bounded domain $G \subset \mathbb{R}^n$ we have

$$\Delta_i^k(h, G)f(x) = \begin{cases} \Delta_i^k(h)f(x) & [x, x + khe_i] \subset G, \\ 0 & [x, x + khe_i] \not\subseteq G \end{cases}$$

Definition 2.8 (Besov spaces) Let $i = 1, 2, 3, m_i, m_0, k_i, k_0 \subset \mathbb{N}$, $m_i > l_i - k_i > 0$, $r \in [1, \infty]$ and we set $l_0, l_i, \in \mathbb{R}_+$, where $l = (l_1, l_2, l_3)$ is the multiindex. Then the Besov space $B_{r,\theta}^{l,l_0}(G)$ is a set of function with finite norm

$$\|u\|_{B_{r,\theta}^{l,l_0}(G)} = \|u\|_{L_r(G)} + \sum_{i=1}^{3} \left(\int_0^{h_0} \left[\frac{\|\Delta_i^{m_i}(h, \Omega)\partial_{x_i}^{k_i} u\|_{L_r(G)}}{h^{l_i - k_i}} \right]^\theta \frac{dh}{h} \right)^{1/\theta}$$

$$+ \left(\int_0^{h_0} \left[\frac{\|\Delta^{m_0}(h, (0, T))\partial_t^{k_0} u\|_{L_r(G)}}{h^{l_0 - k_0}} \right]^\theta \frac{dh}{h} \right)^{1/\theta}, \quad r \in [1, \infty),$$

where h_0 is an arbitrary parameter, $G \subset \Omega \times (0, T)$, and $\theta \in [1, \infty)$. For $r = \infty$ we define

$$\|u\|_{B_{\infty,\theta}^{l,l_0}(G)} = \text{ess} \sup_{(x,t) \in G} |u(x, t)|$$

$$+ \sum_{i=1}^{3} \left(\int_0^{h_0} \left| \frac{\text{ess} \sup_{(x,t) \in G} |\Delta_i^{m_i}(h, \Omega)\partial_{x_i}^{k_i} u|}{h^{l_i - k_i}} \right|^\theta \frac{dh}{h} \right)^{1/\theta}$$

$$+ \left(\int_0^{h_0} \left| \frac{\text{ess} \sup_{(x,t) \in G} |\Delta^{m_0}(h, (0, T))\partial_t^{k_0} u|}{h^{l_0 - k_0}} \right|^\theta \frac{dh}{h} \right)^{1/\theta}.$$

The spaces are equivalent for any m_i, k_i satisfying $m_i + k_i > l_i > k_i > 0$.

Golovkin in [G] proved that the norms of spaces $W_r^{l,l_0}(\Omega^T)$ and $B_{r,\theta}^{l,l_0}(\Omega^T)$ are equivalent.

For $\theta = \infty$, $B_{r,\infty}^{l,l_0}(G)$ are denoted by $H_r^{l,l_0}(G)$ and called the Nikolskii spaces.

Definition 2.9 (Nikolskii spaces) Let $i = 1, 2, 3, m_i, m_0, k_i, k_0 \subset \mathbb{N}$, $m_i > l_i - k_i > 0$, $r \in [1, \infty]$ and we set $l_0, l_i, \in \mathbb{R}_+$. Then the Nikolskii space $H_r^{l,l_0}(G)$ is a set of function with finite norm

$$\|u\|_{H_r^{l,l_0}(G)} = \|u\|_{L_r(G)} + \sum_{i=1}^{3} \sup_{0<h<h_0} \frac{\|\Delta_i^{m_i}(h,\Omega)\partial_{x_i}^{k_i}u\|_{L_r(G)}}{h^{l_i-k_i}}$$

$$+ \sup_{0<h<h_0} \frac{\|\Delta^{m_0}(h,(0,T))\partial_t^{k_0}u\|_{L_r(G)}}{h^{l_0-k_0}}, \quad r \in [1,\infty).$$

For $r = \infty$ we have

$$\|u\|_{H_\infty^{l,l_0}(G)} = \text{ess}\sup_{(x,t)\in G} |u(x,t)| + \sum_{i=1}^{3} \sup_{0<h<h_0} \frac{\text{ess}\sup_{(x,t)\in G} |\Delta_i^{m_i}(h,\Omega)\partial_{x_i}^{k_i}u|}{h^{l_i-k_i}}$$

$$+ \sup_{0<h<h_0} \frac{\text{ess}\sup_{(x,t)\in G} |\Delta^{m_0}(h,(0,T))\partial_t^{k_0}u|}{h^{l_0-k_0}}.$$

Definition 2.10 (Besov space defined for functions on $\Omega \subset \mathbb{R}^n$)
Consider $m_i, k_i \in \mathbb{N}, m_i > l_i - k_i > 0, i = 1,\ldots,n$. Let Ω be an open set in $\mathbb{R}^n, \theta \in [1,\infty]$. Denote by $\bar{l} = (l_1,\ldots,l_n)$ and $\bar{r} = (r_1,\ldots,r_n)$. Then $B_{\bar{r},\theta}^{\bar{l}}(\Omega)$—the Besov space—is a linear normed space of functions u defined on Ω with the norm

$$\|u\|_{B_{\bar{r},\theta}^{\bar{l}}(\Omega)} = \|u\|_{L_{\bar{r}}(\Omega)} + \sum_{i=1}^{n} \left(\int_0^{h_0} \left[\frac{\|\Delta_i^{m_i}(h,\Omega)\partial_{x_i}^{k_i}u\|_{L_{\bar{r}}(\Omega)}}{h^{l_i-k_i}}\right]^\theta \frac{dh}{h}\right)^{1/\theta}$$

and

$$\|u\|_{L_{\bar{r}}(\mathbb{R}^n)} = \left\{\int_{\mathbb{R}} \left[\cdots\left(\int_{\mathbb{R}}\left(\int_{\mathbb{R}}|u(x)|^{r_1}dx_1\right)^{r_2/r_1} dx_2\right)^{r_3/r_2}\cdots\right]^{r_n/r_{n-1}} dx_n\right\}^{1/r_n}.$$

To show the regularity of weak solutions to some elliptic and parabolic problems we need a **partition of unity**. We define two collections of open subsets $\{\omega^{(k)}\}$ and $\{\Omega^{(k)}\}$, $k \in \mathcal{M} \cup \mathcal{N}$, such that

$$\bar{\omega}^{(k)} \subset \Omega^{(k)} \subset \Omega, \bigcup_k \omega^{(k)} = \bigcup_k \Omega^{(k)} = \Omega,$$

$$\bar{\Omega}^{(k)} \cap S = \emptyset \text{ for } k \in \mathcal{M}, \quad \bar{\Omega}^{(k)} \cap S_1 \neq \emptyset \text{ for } k \in \mathcal{N}_1,$$

$$\bar{\Omega}^{(k)} \cap S_2 \neq \emptyset \text{ for } k \in \mathcal{N}_2 \text{ and } \bar{\Omega}^{(k)} \cap L \neq \emptyset \text{ for } k \in \mathcal{N}_3.$$

Hence $\mathcal{N} = \mathcal{N}_1 \cup \mathcal{N}_2 \cup \mathcal{N}_3$ and $\bar{\Omega}^{(k)} \cap S_i$, $k \in \mathcal{N}_i$, $i = 1,2$, are located in a positive distance from the edge $L = \partial S_2$. We assume that at most N_0 of the $\Omega^{(k)}$ have nonempty intersections, and $\sup_k \text{diam } \Omega^{(k)} \leq 2\lambda$,

$\sup_k \operatorname{diam} \omega^{(k)} \leq \lambda$ for some $\lambda > 0$. Let $\zeta^{(k)}(x)$ be a smooth function such that

$$0 \leq \zeta^{(k)}(x) \leq 1, \zeta^{(k)}(x) = 1 \text{ for } x \in \omega^{(k)},$$
$$\zeta^{(k)}(x) = 0 \text{ for } x \in \Omega \setminus \Omega^{(k)}$$

and $|D_x^\nu \zeta^{(k)}(x)| \leq c/\lambda^{|\nu|}$. Then $1 \leq \sum_k (\zeta^{(k)}(x))^2 \leq N_0$. Introducing the function

$$\eta^{(k)}(x) = \frac{\zeta^{(k)}(x)}{\sum_k (\zeta^{(l)}(x))^2},$$

we have that

$$\eta^{(k)}(x) = 0 \text{ for } x \in \Omega \setminus \Omega^{(k)}, \sum_k \eta^{(k)}(x)\zeta^{(k)}(x) = 1 \text{ and } |D_x^\nu \eta^{(k)}(x)| \leq c/\lambda^{|\nu|}.$$

We also by φ denote an increasing positive function of its arguments.

2.2 Auxiliary Problems, Results, and Green Functions

Consider the Neumann problem for **the Poisson equation**

$$-\Delta u = f \quad \text{in} \quad \Omega,$$
$$n \cdot \nabla u = \alpha \quad \text{on} \quad S, \tag{2.4}$$

where $S = S_1 \cup S_2, \Omega$ is the cylindrical domain with the dihedral angle of $\pi/2$ between S_1 and S_2. Moreover, the following compatibility condition holds

$$\int_\Omega f dx = \int_S \alpha dS. \tag{2.5}$$

Lemma 2.11 *Let f, α be so regular that $f \in L_2(\Omega), \alpha_i \in H^{1/2}(S_i), i = 1, 2$. Let u be a solution to problem (2.4). Then there exists a Green function to problem (2.4) such that any solution to (2.4) can be expressed in the form*

$$u(x) = \int_\Omega G(x, y) f(y) dy - \int_S \alpha(s) G(x, s) dS(s). \tag{2.6}$$

Assume that $f \in L_2(\Omega), \alpha_i \in H^{1/2}(S_i), i = 1, 2$. Then there exists a solution to (2.4) such that $u \in H^2(\Omega)$ and

$$\|u\|_{2,\Omega} \le c(|f|_{2,\Omega} + \sum_{i=1}^{2} \|\alpha_i\|_{1/2,S_i}). \qquad (2.7)$$

Proof By the Green function to (2.4) we mean a solution to the problem

$$
\begin{aligned}
-\Delta_x G(x,y) &= \delta(x-y) \text{ in } \Omega, \\
n \cdot \nabla_x G(x,y) &= 0 \text{ on } S,
\end{aligned}
\qquad (2.8)
$$

where $\delta(x-y)$ is the delta Dirac function and y is an arbitrary parameter. We are looking for solutions to (2.8) in the form

$$G(x,y) = \sum_{k \in \mathcal{M} \bigcup \mathcal{N}_1 \bigcup \mathcal{N}_2 \bigcup \mathcal{N}_3} \eta_k(x,y) E_k(x,y) \zeta^{(k)}(x) + g(x,y), \qquad (2.9)$$

where $\{\zeta^{(k)}(x)\}_{k \in \mathcal{M} \bigcup \mathcal{N}_1 \bigcup \mathcal{N}_2 \bigcup \mathcal{N}_3}$ is the partition of unity that we introduce in this chapter, and $\eta_k(x,y)$ is a smooth function such that

$$\eta_k(x,y) = \begin{cases} 1 & |x-y| \le 1 \\ 0 & |x-y| \ge 2, \end{cases}$$

where $x \in \operatorname{supp} \eta^{(k)}$.

For $k \in \mathcal{M}$ we have that

$$E_k(x,y) = \frac{1}{4\pi} \frac{1}{|x-y|}.$$

Let $k \in \mathcal{N}_1$. Then we set $S_{1k} = \operatorname{supp} \eta^{(k)} \bigcap S_1$ and introduce such local coordinates (x,y) that $S_{1k} = \{x : x_3 = 0\}$. Then $E_k(x,y)$ is the Green function to the Neumann problem in $\mathbb{R}^3_+$. Hence

$$E_k(x,y) = \frac{1}{4\pi} \left(\frac{1}{|x-y|} + \frac{1}{|x-\bar{y}|} \right),$$

where $\bar{y} = (y_1, y_2, -y_3)$.

For $k \in \mathcal{N}_2$ and for $\operatorname{supp} \eta^{(k)}$ located in a positive distance from L, $E_k(x,y)$ is also the Green function to the Neumann problem in $\mathbb{R}^3_+$.

For $k \in \mathcal{N}_3$ we introduce such coordinates that L becomes locally $x_2 = x_3 = 0$ and then in these coordinates $E_k(x,y)$ equals to

$$E_k(x,y) = \frac{1}{4\pi} \left(\frac{1}{|x-y|} + \frac{1}{|x-\bar{y}_1|} + \frac{1}{|x-\bar{y}_2|} + \frac{1}{|x-\bar{y}_3|} \right),$$

where $\bar{y}_1 = (y_1, -y_2, y_3), \bar{y}_2 = (y_1, y_2, -y_3)$, and $\bar{y}_3 = (y_1, -y_2, -y_3)$.

Inserting (2.9) in (2.8) we derive the following problem for g

$$-\Delta g = \sum_k (2\nabla E_k \nabla(\eta_k \zeta^{(k)}) + E_k \Delta(\eta_k \zeta^{(k)})) \equiv G \quad \text{in} \ \Omega,$$

$$\frac{\partial g}{\partial n} = \sum_k \bar{n} \cdot \nabla E_k \eta_k \zeta^{(k)} \equiv H \quad \text{on} \ S, \qquad (2.10)$$

$$\int_\Omega g dx = 0.$$

In order to prove the existence of solutions to problem (2.10) we use the partition of unity and the technique of regularizer (see [Z8], Lemma 2.3). The technique of regularizer for elliptic equations can be found in the papers of Solonnikov [S1] and Agmon et al. [ADN]. However, to demonstrate the existence we also need the estimate for weak solutions to problem (2.10) of the form

$$\|g\|_{1,\Omega} \leq c(|G|_{2,\Omega} + |H|_{2,S}). \qquad (2.11)$$

Applying this, the existence of the Green function satisfying (2.8) is established. To show (2.7) we use the Calderón-Zygmund estimates for singular integrals (see the book of Stein [St]). We have to emphasize that only the first term in $E_k(x,y)$ is the singular kernel when others are not. Thus the Stein results are necessary to estimate the singular part only. The others parts of $E_k(x,y)$ can be easily bounded with (2.7). This concludes the proof. □

Now, we consider the Dirichlet problem for **the stationary Stokes system**

$$-\nu\Delta u + \nabla p = f \qquad\qquad \text{in} \ \Omega$$

$$\text{div} \, u = 0 \qquad\qquad \text{in} \ \Omega$$

$$u|_{S_i} = \alpha_i, \ i = 1,2 \quad \text{on} \ S = S_1 \cup S_2, \qquad (2.12)$$

$$\int_\Omega p \, dx = 0,$$

where Ω is the cylindrical domain and we assume the right dihedral angle between S_1 and S_2. For problem (2.12) we need the following compatibility condition

$$\sum_{i=1}^2 \int_{S_i} \alpha_i \cdot \bar{n}_i dS_i = 0, \qquad (2.13)$$

where $\bar{n}_i$ is the unit exterior vector normal to $S_i, i = 1,2$.

Lemma 2.12 *There exists the Green function*

$$G_{ij}(x,y), g_j(x,y), \quad i,j = 1,2,3,$$

to problem (2.12) such that

$$\nu \nabla_x G_{ij}(x,y) + \partial_{x_i} g_j(x,y) = \delta(x-y)e_{ij},$$
$$\operatorname{div}_x G_{ij}(x,y) = 0, \qquad (2.14)$$
$$G_{ij}(x,y) = 0, \quad x,y \in S_\sigma, \ \sigma = 1,2,$$

where $\delta(x-y)$ is the Dirac delta function, and $e_j = (e_{1j}, e_{2j}, e_{3j})$ is the unit vector directed along the j-th coordinate axis. Moreover, any solution to (2.12) is expressed in the form

$$u_i(x) = \int_\Omega G_{ij}(x,y) f_j(y) dy + \sum_{\sigma=1}^2 \int_{S_\sigma} \alpha_{j\sigma}(G_{ji,x_l} + G_{li,x_j})\bar{n}_{\sigma l} dS_\sigma, \quad (2.15)$$

where the summation convention over twice-repeated indices is assumed.

Proof To prove the existence of the Green function we construct it in the following form

$$G_{ij}(x,y) = \sum_{k \in \mathcal{M} \cup \mathcal{N}_1 \cup \mathcal{N}_2 \cup \mathcal{N}_3} \eta_k(x,y) U_{ij}^{(k)}(x,y) \zeta^{(k)}(x) + H_{ij}(x,y),$$
$$\qquad (2.16)$$
$$g_j(x,y) = \sum_{k \in \mathcal{M} \cup \mathcal{N}_1 \cup \mathcal{N}_2 \cup \mathcal{N}_3} \eta_k(x,y) Q_j^{(k)}(x,y) \zeta^{(k)}(x) + K_j(x,y),$$

where $\{\zeta^{(k)}(x)\}_{k \in \mathcal{M} \cup \mathcal{N}_1 \cup \mathcal{N}_2 \cup \mathcal{N}_3}$ is the partition of unity introduced in this chapter and $\eta_k(x,y)$ is a smooth function such that

$$\eta_k(x,y) = \begin{cases} 1 & |x-y| \le 1 \\ 0 & |x-y| \ge 2, \end{cases}$$

where $x \in \operatorname{supp} \eta^{(k)}$. For $k \in \mathcal{M}$ we have (see the books of Ladyzhenskaya [L1, Chapter 3,(5)], and Galdi [G1, Chapter 4, Sect. 2, (2.23)])

$$U_{ij}^{(k)}(x,y) = -\frac{1}{8\pi\nu}\left[\frac{\delta_{ij}}{|x-y|} + \frac{(x_j-y_j)(x_i-y_i)}{|x-y|^3}\right],$$
$$Q_j^{(k)}(x,y) = -\frac{x_j-y_j}{4\pi|x-y|^3}.$$

For $k \in \mathcal{N}_1$ we introduce new coordinates in $\operatorname{supp} \zeta^{(k)}$ such that transformed $S_1 \cap \operatorname{supp} \zeta^{(k)} = \{x : x_3 = 0\}$. In these new coordinates we have the fundamental solution in $\mathbb{R}^3_+$ in the form (3.36) in [G1, Ch. 4, Sect. 3]. The fundamental solution satisfies

$$\nu \Delta_x U_{ij}^{(k)}(x, y) + \nabla_{x_i} Q_j^{(k)} = \delta(x - y)\delta_{ij},$$

$$\operatorname{div}_{x_i} U_{ij}^{(k)}(x, y) = 0, \qquad (2.17)$$

$$U_{ij}^{(k)}(x, y) = 0, \quad \text{for } x, y \in \{x : x_3 = 0\}.$$

For $k \in \mathcal{N}_2$ we have fundamental solutions—the same as described by (2.17). Thankfully, however, in this case we do not need to introduce new coordinates because S_2 is flat. We only require that $\operatorname{supp} \zeta^{(k)}$ must be located in a positive distance from L.

Finally, we choose $k \in \mathcal{N}_3$, so we consider neighborhoods of L. Take an arbitrary point $p_0 \in L$. Then we take $\zeta^{(k)}$ such that $\operatorname{supp} \zeta^{(k)}$ is a neighborhood of p_0. Since S_1 and S_2 meet along L and S_2 is flat we introduce new local coordinates with origin at p_0 such that $S_1 \cap \operatorname{supp} \zeta^{(k)}$ becomes flat, too. Moreover, we require that coordinates $x = (x_1, x_2, x_3)$ are such that:

$$L = \{x : x_2 = x_3 = 0\}, S_1 \cap \operatorname{supp} \zeta^{(k)} \subset \{x : x_2 = 0\}, S_2 \subset \{x : x_3 = 0\}.$$

Since we consider the nonhomogeneous Dirichlet boundary problem we make it homogeneous on S_1 by an appropriate extension. Then we extend the problem by reflection with respect to transformed $S_1 \cap \operatorname{supp} \zeta^{(k)}$. In this way we derive the problem in the half space $\mathbb{R}^3_+ = \{x : x_3 > 0\}$. In this case we have the fundamental solution in the form described in (2.17). Thus, we conclude that the first terms on the r.h.s. of (2.16) can be constructed after passing to old coordinates.

Inserting (2.16) in (2.14) we obtain the problem

$$-\nu \Delta H_{ij} + \nabla_{x_i} K_j = -\sum_k [2\nabla U_{ij}^{(k)} \nabla(\eta_k \zeta^{(k)})$$

$$+ U_{ij}^{(k)} \Delta(\eta_k \zeta^{(k)}) + Q_j^{(k)} \nabla_{x_i}(\eta_k \zeta^{(k)})] \quad \text{in } \mathbb{R}^3_+,$$

$$\operatorname{div}_{x_i} H_{ij}(x, y) = 0 \quad \text{in } \mathbb{R}^3_+, \qquad (2.18)$$

$$H_{ij} = -\sum_k \eta_k U_{ij}^{(k)} \zeta^{(k)} \quad \text{on } S.$$

The existence of solutions to (2.18) can be then proved by the method of regularizer and the Fredholm theorem (see [ADN], [S1]). This is shown in Lemma 2.13 below. This concludes the proof. $\qquad \square$

Lemma 2.13 *Let (u, p) be a solution to the stationary Stokes system (2.12).*
Assume that $f \in L_2(\Omega), \alpha_i \in H^{3/2}(S_i), i = 1, 2$, and Ω, S_1, S_2 are described
in the formulation of problem (2.12) and $S_1 \in C^2$. Then there exists a unique
solution to problem (2.12) such that $u \in H^2(\Omega), p \in H^1(\Omega)$ and the following
estimate holds

$$\|u\|_{H^2(\Omega)} + \|p\|_{H^1(\Omega)} \leq c(\|f\|_{L_2(\Omega)} + \sum_{i=1}^{2} \|\alpha_i\|_{H^{3/2}(S_i)}). \qquad (2.19)$$

Proof Repeating the proof of Lemma 9.4 (see Chap. 9) we obtain for
solutions to problem (2.12) the following inequality

$$\|u\|_{H^2(\Omega)} + \|p\|_{H^1(\Omega)}$$

$$\leq c(\|f\|_{L_2(\Omega)} + \sum_{i=1}^{2} \|\alpha_i\|_{H^{3/2}(S_i)} + \|u\|_{L_2(\Omega)} + \|p\|_{L_2(\Omega)}). \qquad (2.20)$$

Let w be a divergence free extension of boundary data such that

$$w|_{S_i} = \alpha_i, \quad i = 1, 2, \qquad (2.21)$$

and

$$\|w\|_{H^2(\Omega)} \leq c \sum_{i=1}^{2} \|\alpha_i\|_{H^{3/2}(S_i)}. \qquad (2.22)$$

Then we introduce the function

$$v = u - w \qquad (2.23)$$

and (v, p) is a solution to the problem

$$\begin{aligned}
-\nu \Delta v + \nabla p &= f + \nu \Delta w & \text{in } \Omega, \\
\operatorname{div} v &= 0 & \text{in } \Omega, \\
v &= 0 & \text{on } S_i, \ i = 1, 2, \\
\int_\Omega p \, dx &= 0.
\end{aligned} \qquad (2.24)$$

Multiplying (2.24) by v and integrating over Ω yield

$$\|v\|_{H^1(\Omega)} \leq c(\|f\|_{L_2(\Omega)} + \|\Delta w\|_{L_2(\Omega)}), \qquad (2.25)$$

where the Poincaré inequality was used.

We introduce function φ such that

$$
\begin{aligned}
\operatorname{div} \varphi &= p, && \text{in } \Omega, \\
\varphi &= 0 && \text{on } S_i, \ i = 1, 2.
\end{aligned}
\tag{2.26}
$$

Then from Kapitanskii and Pileckas [KP] it follows the existence of $\varphi \in H^1(\Omega)$ and the estimate

$$
\|\varphi\|_{H^1(\Omega)} \leq c\|p\|_{L_2(\Omega)}.
\tag{2.27}
$$

Multiplying $(2.24)_1$ by φ, integrating over Ω, and using the properties of φ we deduce

$$
\|p\|_{L_2(\Omega)} \leq c(\|f\|_{L_2(\Omega)} + \|\Delta w\|_{L_2(\Omega)}).
\tag{2.28}
$$

Applying (2.25) and (2.28) in (2.20)

$$
\|u\|_{H^2(\Omega)} + \|p\|_{H^1(\Omega)} \leq c(\|f\|_{L_2(\Omega)} + \sum_{i=1}^{2} \|\alpha_i\|_{H^{3/2}(S_i)}).
\tag{2.29}
$$

Hence (2.29) implies uniqueness. Then the Fredholm theorem also gives the existence. This concludes the proof. $\square$

Consider the nonstationary Stokes system

$$
\begin{aligned}
u_t - \nu \Delta u + \nabla p &= f && \text{in } \Omega^T, \\
\operatorname{div} u &= 0 && \text{in } \Omega^T, \\
u &= d && \text{on } S^T, \\
u &= u(0) && \text{in } \Omega.
\end{aligned}
\tag{2.30}
$$

Recall Theorem A from [Z8] (see also [NZ2]).

Lemma 2.14 *Assume that* $f \in L_2(\Omega^T)$, $d_i \in W_\sigma^{2-1/\sigma,1-1/2\sigma}(S_i^T)$, $i = 1, 2$, *and* $u(0) \in W_\sigma^{2-2/\sigma}(\Omega)$ $\sigma \in (1, \infty)$. *Then there exists a solution to (2.30) such that* $u \in W_\sigma^{2,1}(\Omega^T)$, $\nabla p \in L_\sigma(\Omega^T)$ *and the estimate holds*

$$
\|u\|_{W_\sigma^{2,1}(\Omega^T)} + \|\nabla p\|_{L_\sigma(\Omega^T)}
$$

$$
\leq c\left(\|f\|_{L_\sigma(\Omega^T)} + \sum_{i=1}^{2} \|d_i\|_{W_\sigma^{2-1/\sigma,1-1/2\sigma}(S_i^T)} + \|u(0)\|_{W_\sigma^{2-2/\sigma}(\Omega)}\right).
\tag{2.31}
$$

2.3   Interpolation, Imbeddings, Trace Theorems, and the Korn Inequality

We need the interpolation inequality.

Lemma 2.15 (see [BIN, Ch. 3, Sect. 15]) *The following inequality holds*

$$\sum_{|\alpha|=r} |D^\alpha f|_{p,\Omega} \le c\|f\|_{l,p_2,\Omega}^\theta |f|_{p_1,\Omega}^{1-\theta}, \quad \Omega \subset \mathbb{R}^3, \tag{2.32}$$

where parameters $p_1, p_2 \in [1,\infty]$, $0 \le r < l$ satisfy

$$\frac{3}{p} - r = (1-\theta)\frac{3}{p_1} + \theta\left(\frac{3}{p_2} - l\right), \quad \frac{r}{l} \le \theta \le 1. \tag{2.33}$$

Now we recall some theorems of traces and extensions for functions defined in $\mathbb{R}^n \times \mathbb{R}_+$, where $\Omega \subset \mathbb{R}^n, (0,T) \subset \mathbb{R}_+$, proved by Bugrov in [B] and Solonnikov in [S2]. Applying a partition of unity and using appropriate smoothness of the boundary, the same theorems hold for Ω^T.

Lemma 2.16 (Trace Theorem) *(see [B], [S2], [FVSU]) Assume that $u \in W_r^{l,l/2}(\mathbb{R}^n \times \mathbb{R}_+)$, $r \in (1,\infty), l \in \mathbb{N}_0$. Let $\tilde{u} = u|_{x_n=0}$. Then $\tilde{u} \in W_r^{l-1/r,l/2-1/2r}(\mathbb{R}^{n-1} \times \mathbb{R}_+)$ and*

$$\|\tilde{u}\|_{W_r^{l-1/r,l/2-1/2r}(\mathbb{R}^{n-1}\times\mathbb{R}_+)} \le c\|u\|_{W_r^{l,l/2}(\mathbb{R}^n\times\mathbb{R}_+)}. \tag{2.34}$$

Lemma 2.17 (Inverse Trace Theorem) *(see [B], [S2], [FVSU]) Assume that $u_j \in W_r^{\sigma-j-1/r,\sigma/2-j/2-1/2r}(\mathbb{R}^{n-1} \times \mathbb{R}_+)$, where $j = 0,1,\ldots,\sigma_*$ and $\sigma_* = [\sigma]$ for $\sigma \notin \mathbb{N}, \sigma > 1/r$ and $\sigma_* = \sigma - 1$ for $\sigma \in \mathbb{N}$. Then there exists a function $u \in W_r^{\sigma,\sigma/2}(\mathbb{R}^n \times \mathbb{R}_+)$ such that*

$$\frac{\partial^j u}{\partial x_n^j} = u_j, \quad j = 0,\ldots,\sigma_*$$

and

$$\|u\|_{W_r^{\sigma,\sigma/2}(\mathbb{R}^n\times\mathbb{R}_+)} \le c\sum_{j=0}^{\sigma_*} \|u_j\|_{W_r^{\sigma-j-1/r,\sigma/2-j/2-1/2r}(\mathbb{R}^{n-1}\times\mathbb{R}_+)}. \tag{2.35}$$

Lemma 2.18 *(see [B], [S2]) Assume that $u \in W_r^{l,l/2}(\mathbb{R}^n \times \mathbb{R}_+)$, where the Cartesian coordinates are such that $x = (x_1,\ldots,x_n) \in \mathbb{R}^n, x_0 \in \mathbb{R}_+, l \in \mathbb{N}, r \in (1,\infty)$. Then $u|_{x_0=0} = \tilde{u} \in W_r^{l-2/r}(\mathbb{R}^n)$ and*

$$\|\tilde{u}\|_{W_r^{l-2/r}(\mathbb{R}^n)} \leq c\|u\|_{W_r^{l,l/2}(\mathbb{R}^n \times \mathbb{R}_+)}. \tag{2.36}$$

Lemma 2.19 *(see [B], [S2]) Assume that* $u \in W_r^\sigma(\mathbb{R}^n)$, $r \in (1,\infty)$, $\sigma \in \mathbb{R}_+$. *Then there exists an extension* $\tilde{u} \in W_r^{\sigma+2/r,\sigma/2+1/r}(\mathbb{R}^n \times \mathbb{R}_+)$ *such that* $\tilde{u}|_{x_0=0} = u$ *and*

$$\|\tilde{u}\|_{W_r^{\sigma+2/r,\sigma/2+1/r}(\mathbb{R}^n \times \mathbb{R}_+)} \leq c\|u\|_{W_r^\sigma(\mathbb{R}^n)}. \tag{2.37}$$

Lemma 2.20 ([BIN, Theorem 18.10]) *Let* $\Omega \subset \mathbb{R}^n$ *be an open set. Let*

$$1 \leq p_i \leq q_i \leq \infty, \qquad i = 1,\ldots,n \qquad \sum_{i=1}^n \left(\frac{1}{p_i} - \frac{1}{q_i} + \alpha_i\right)\frac{1}{l_i} \leq 1.$$

Let either $\theta = 1$ *or* $1 \leq \theta \leq q_n$, $1 \leq p_n \leq q_n < \infty$. *Then*

$$D^\alpha B_{\bar{p},\theta}^{\bar{l}}(\Omega) \subset L_{\bar{q}}(\Omega)$$

where $\bar{l} = (l_1,\ldots,l_n), \bar{p} = (p_1,\ldots,p_n), \bar{q} = (q_1,\ldots,q_n)$, *and*

$$\|D_x^\alpha u\|_{L_{\bar{q}}(\Omega)} \leq c\|u\|_{B_{\bar{p},\theta}^{\bar{l}}(\Omega)}, \tag{2.38}$$

where $\alpha = (\alpha_1,\ldots,\alpha_n)$ *is the multiindex and*

$$D_x^\alpha = \partial_{x_1}^{\alpha_1}\ldots\partial_{x_n}^{\alpha_n}, \quad \alpha_i \in \mathbb{N}_0, i = 1,\ldots,n.$$

Lemma 2.21 (see the Proof of Lemma 3.7 in [Z3]) *Assume that* $v \in V^1(\Omega^T)$. *Then* $\nabla v \in L_{10/3}(\Omega^T), v \in L_{10}(\Omega^T)$ *and there exists a constant* c *such that*

$$\|\nabla v\|_{L_{10/3}(\Omega^T)} \leq c\|v\|_{V^1(\Omega^T)}, \tag{2.39}$$

$$\|v\|_{L_{10}(\Omega^T)} \leq c\|v\|_{V^1(\Omega^T)}. \tag{2.40}$$

Proof If $v \in V^1(\Omega^T)$, then $\nabla v \in V^0(\Omega^T)$. Hence

$$\|\nabla v\|_{L_q(0,T;L_p(\Omega))} \leq c\|\nabla v\|_{V^0(\Omega^T)}, \tag{2.41}$$

where

$$\frac{3}{p} + \frac{2}{q} = \frac{3}{2}. \tag{2.42}$$

For $p = q = \frac{10}{3}$ we derive (2.39).

If $v \in V^1(\Omega^T)$, then $v, \nabla v \in L_q(0, T; L_p(\Omega))$ where p, q satisfy (2.42). We apply the imbedding (see [BIN, Ch. 3, Sect. 10], and also the book of Adams [Ad]).

$$W_p^1(\Omega) \subset L_\sigma(\Omega) \quad \text{where} \quad \frac{3}{p} - \frac{3}{\sigma} \leq 1.$$

Then $v \in L_q(0, T; L_\sigma(\Omega))$ and $p = \dfrac{3\sigma}{3+\sigma}$. Inserting the expression for p into (2.42) we get

$$\frac{3}{\sigma} + \frac{2}{q} = \frac{1}{2}.$$

Then for $\sigma = q$ we get that $q = 10$. This implies (2.38) and concludes the proof. $\square$

We need also the following Korn inequality

Lemma 2.22 *Assume that $E_\Omega(w) = |\mathbb{D}(w)|_{2,\Omega}^2 < \infty, \operatorname{div} w = 0, w \cdot n|_S = 0$, where Ω is not axially symmetric. Then*

$$\|w\|_{1,\Omega}^2 \leq c E_\Omega(w). \tag{2.43}$$

Proof

$$|\mathbb{D}(w)|_{2,\Omega}^2 = \int_\Omega \left(\frac{\partial w_i}{\partial x_j} + \frac{\partial w_j}{\partial x_i} \right)^2 dx = \int_\Omega \left(\left| \frac{\partial w_i}{\partial x_j} \right|^2 + \left| \frac{\partial w_j}{\partial x_i} \right|^2 \right) dx$$

$$+ 2 \int_\Omega \frac{\partial w_i}{\partial x_j} \frac{\partial w_j}{\partial x_i} dx \equiv I_1 + 2 I_2.$$

Integrating by parts in the last integral and using that w is divergence free we have

$$I_2 = \int_\Omega \partial_{x_i} \left(\frac{\partial w_i}{\partial x_j} w_j \right) dx = \int_S n_i \frac{\partial w_i}{\partial x_j} w_j dS = - \int_S n_{i,x_j} \cdot w_i w_j dS,$$

where we used that $w \cdot \bar{n} = 0$. Then we obtain

$$|\nabla w|_{2,\Omega}^2 \leq c E_\Omega(w) + c |w|_{2,S}^2. \tag{2.44}$$

Next by the contradiction argument (see Solonnikov and Shchadilov [SS] and the proof of Lemma 9.2) we get

$$|w|_{2,\Omega}^2 \leq \delta |\nabla w|_{2,\Omega}^2 + M E_\Omega(w), \tag{2.45}$$

where δ can be chosen sufficiently small. Hence (2.44), (2.45) imply (2.43) and we conclude the proof. $\qquad\square$

Remark 2.23 Assume that $E_\Omega(w) = |\mathbb{D}(w)|^2_{2,\Omega} < \infty, \operatorname{div} w = 0$, and

$$w \cdot \bar{n}|_{S_2} = 0, \quad \bar{n} \cdot \mathbb{D}(w) \cdot \bar{e}_r|_{S_2} = 0, \quad w_\varphi|_{S_2} = \beta, \tag{2.46}$$

where β is a given function and Ω is axially symmetric. Then

$$\|w\|^2_{1,\Omega} \le c(E_\Omega(w) + |\beta|^2_{2,S_2} + \alpha^2_*), \tag{2.47}$$

where α_* is introduced in (2.50).

Proof Introduce the function w defined by (3.7) which is a solution to (3.8). Let

$$\eta = (-x_2, x_1, 0) \quad \text{and} \quad rv_\varphi = v \cdot \eta.$$

Multiply $(3.8)_1$ by η and integrate over Ω. Then we have

$$\frac{d}{dt} \int_\Omega w \cdot \eta dx + \int_\Omega w \cdot \nabla w \cdot \eta dx + \int_\Omega w \cdot \nabla \delta \cdot \eta dx + \int_\Omega \delta \cdot \nabla w \cdot \eta dx$$
$$- \int_\Omega \operatorname{div} \mathbb{T}(w, p) \cdot \eta dx = \int_\Omega F(\delta, t) \cdot \eta dx. \tag{2.48}$$

Integrating by parts in the second term on the l.h.s. of (2.48) and using that $\operatorname{div} w = 0$, $w \cdot \bar{n}|_S = 0$ we obtain

$$\int_\Omega w \cdot \nabla w \cdot \eta dx = -\int_\Omega w_i w_j \eta_{j,x_i} dx = 0$$

because we have the trace of product of symmetric and antisymmetric tensors.

Consider the sum of the third and the fourth terms on the l.h.s. of (2.48). Integrating by parts and using that $\operatorname{div} w = 0, \operatorname{div} \delta = 0, w \cdot \bar{n}|_S = 0$ we obtain

$$\int_\Omega w \cdot \nabla \delta \cdot \eta dx + \int_\Omega \delta \cdot \nabla w \cdot \eta dx = -\int_\Omega (w_i \delta_j + w_j \delta_i) \eta_{j,x_i} dx - \int_S \delta \cdot \bar{n} w \cdot \eta dS.$$

Integrating by parts in the last term on the l.h.s. of (2.48) we have

$$-\int_\Omega \operatorname{div} \mathbb{T}(w, p) \cdot \eta dx = -\int_S \bar{n} \cdot \operatorname{div} \mathbb{T}(w, p) \cdot \eta dS + \int_\Omega \operatorname{div} \mathbb{T}(w, p) \cdot \nabla \eta dx$$
$$= -\int_{S_1} B_{1\alpha}(\delta)\eta \cdot \bar{\tau}_\alpha dS_1 - \int_{S_2} B_{2\alpha}(\delta)\eta \cdot \bar{\tau}_\alpha dS_2 \equiv I,$$

where we used that $\operatorname{div} \mathbb{T}(w, p) \cdot \nabla \eta = 0$ because of the trace of product of symmetric and antisymmetric tensors. Recall that $\bar{\tau}_1 = \frac{\eta}{|\eta|}$ and $\bar{\tau}_3 = (0, 0, 1)$, thus we calculate

$$I = -\int_{S_1} B_{11}(\delta)|\eta|dS_1 - \int_{S_2} B_{21}(\delta)|\eta|dS_2.$$

Summarizing the considerations on terms in (2.48) yields

$$\frac{d}{dt}\int_{\Omega} w \cdot \eta dx = \int_{S_2} \alpha w \cdot \eta dS_2 + \int_{S_1} B_{11}(\delta)|\eta|dS_1$$
$$+ \int_{S_2} B_{21}(\delta)|\eta|dS_2 + \int_{\Omega} F(\delta, t) \cdot \eta dx. \tag{2.49}$$

Integrating (2.49) with respect to time implies

$$\int_{\Omega} w \cdot \eta dx \le \int_{S_2^t} \alpha w \cdot \eta dS_2 dt' + \int_{S_1^t} B_{11}(\delta)|\eta|dS_1 dt'$$
$$+ \int_{S_2^t} B_{21}(\delta)|\eta|dS_2 dt' + \int_{\Omega^t} F(\delta, t') \cdot \eta dx dt' + \int_{\Omega} w(0) \cdot \eta dx \equiv \alpha_*. \tag{2.50}$$

We can write w in the form

$$w = w' + \frac{\alpha_*}{|\eta|_{2,\Omega}^2}\eta, \tag{2.51}$$

where

$$w' = w_r \bar{e}_r + w_\varphi \bar{e}_\varphi + w_z \bar{e}_z - \frac{\alpha_*}{|\eta|_{2,\Omega}^2}\eta$$
$$= w_r \bar{e}_r + \left(w_\varphi - \frac{\alpha_*}{|\eta|_{2,\Omega}^2}r\right)\bar{e}_\varphi + w_z \bar{e}_z \tag{2.52}$$

and

$$\int_{\Omega} w' \cdot \eta dx = 0 \tag{2.53}$$

Then in view of Lemma 4.4 from [SS] it follows that

$$|\nabla w'|_{2,\Omega}^2 \le cE_{\Omega}(w'). \tag{2.54}$$

Since $E_\Omega(\eta) = 0$ Eqs. (2.51) and (2.54) imply

$$|\nabla w|_{2,\Omega}^2 \leq c(E_\Omega(w) + |\int_\Omega w \cdot \eta dx|^2). \tag{2.55}$$

From $w_r|_{S_1} = 0, w_z|_{S_2} = 0$ and the Poincaré inequality we have

$$\begin{aligned}
|w_r|_{2,\Omega} &\leq c|w_{r,r}|_{2,\Omega} \leq c|\nabla w|_{2,\Omega}, \\
|w_z|_{2,\Omega} &\leq c|w_{z,z}|_{2,\Omega} \leq c|\nabla w|_{2,\Omega}.
\end{aligned} \tag{2.56}$$

To obtain an estimate for $w_\varphi = w \cdot \bar{e}_\varphi$ we need different boundary conditions on S_2, i.e., (2.46)

$$w \cdot \bar{n}|_{S_2} = 0, \quad \bar{n} \cdot \mathbb{D}(w) \cdot \bar{e}_r|_{S_2} = 0, \quad w_\varphi|_{S_2} = \beta,$$

where β is a given function. Then

$$|w_\varphi|_{2,\Omega} \leq c|\nabla w_\varphi|_{2,\Omega} + c|\beta|_{2,S_2}. \tag{2.57}$$

To derive the estimates we introduce the function

$$w'_\varphi = w_\varphi - \tilde{\beta},$$

where $\tilde{\beta}$ is an extension of β such that $\tilde{\beta}|_{S_2} = \beta$. Collecting the estimates yields (2.47). $\qquad\square$

The new boundary conditions would have a strong influence on the proof presented in the book. This is why we consider the non-axially symmetric domain Ω.

Recall Lemma 18.12 from [BIN, Ch. 4, Sect. 18].

Lemma 2.24 *Assume that* $u \in W_\sigma^{l,l/2}(\Omega^T)$, $\sigma \in (1,\infty)$, $l \in \mathbb{N}$. *Assume also*

$$\left(\frac{3}{\sigma} - \frac{3}{p}\right)\frac{1}{l} + \left(\frac{1}{\sigma} - \frac{1}{q}\right)\frac{2}{l} + l'\frac{2}{l} \leq 1,$$

$$\mu_0 = \frac{3}{p_0}\frac{1}{l} + \frac{1}{q_0}\frac{2}{l} + l'\frac{2}{l},$$

$$\mu_1 = 1 - \left(\frac{3}{\sigma} - \frac{3}{p}\right)\frac{1}{l} + \left(\frac{1}{\sigma} - \frac{1}{q}\right)\frac{2}{l} + l'\frac{2}{l} > 0,$$

$$p, q \in (1,\infty), l' \in (0,1).$$

Then $u \in L_p(\Omega; B_{q,\theta}^{l'}(0,T))$ and there exists a constant c such that

$$\|u\|_{L_p(\Omega; B_{q,\theta}^{l'}(0,T))} \leq c\varepsilon^{-\mu_0} \|u\|_{L_{q_0}(0,T; L_{p_0}(\Omega))} + c\varepsilon^{\mu_1} \|u\|_{W_\sigma^{l,l/2}(\Omega^T)}. \quad (2.58)$$

Recall Lemma 10.1 from [BIN, Ch. 3, Sect. 10] on the anisotropic Sobolev space with the mixed norm: $W_{p_1,p_2}^{l,l/2}(\Omega^T)$, where l-even, $p_1, p_2 \in [1, \infty]$.

Lemma 2.25 *Assume that $u \in W_{p_1,p_2}^{l,l/2}(\Omega^T)$, l-even, $p_1, p_2 \in [1, \infty]$. Let $q_i, i = 1, 2$ be such that $1 \leq p_i \leq q_i \leq \infty, i = 1, 2$. If*

$$\varkappa = \left(\frac{3}{p_1} - \frac{3}{q_1}\right)\frac{1}{l} + \left(\frac{1}{p_2} - \frac{1}{q_2}\right)\frac{2}{l} + \sum_{i=0}^{3} \alpha_i \frac{1}{l_i} < 1,$$

where $l_0 = l/2, l_1 = l_2 = l_3 = l$, $\alpha_i \in \mathbb{N}_0, i = 0, 1, 2, 3$ and α is multiindex, then, with $D^\alpha = \partial_t^{\alpha_0} \partial_{x_1}^{\alpha_1} \partial_{x_2}^{\alpha_2} \partial_{x_3}^{\alpha_3}$, $D^\alpha u \in L_{q_2}(0,T; L_{q_1}(\Omega))$,

$$\|D^\alpha u\|_{L_{q_2}(0,T; L_{q_1}(\Omega))} \leq c\varepsilon^{1-\varkappa} \|u\|_{W_{p_1,p_2}^{l,l/2}(\Omega^T)} + c\varepsilon^{-\varkappa} \|u\|_{L_{q_2}(0,T; L_{q_1}(\Omega))}. \quad (2.59)$$

Chapter 3
Energy Estimate: Global Weak Solutions

Abstract In this chapter we derive the energy estimate for w—a weak solution to the system with homogeneous Dirichlet boundary conditions on $S-$ the boundary of domain Ω. To construct such solutions we use Hopf function and then an extension δ such that $w = v - \delta$ satisfies $w \cdot \bar{n}|_S = 0$ so in equations for w it is possible to integrate by parts. However, to obtain an energy estimate we need some refined estimates in weighted Sobolev spaces for the extension δ and for solutions to the auxiliary Neumann problem for the Poisson equation—these are shown in Chaps. 10 and 11.

Further, in Proposition 3.4 we prove that

$$\|v\|_{V(kT,t)} \leq \mathcal{A}, \quad t \in [kT, (k+1)T], \quad k \in \mathbb{N}_0, \quad \text{where}$$

$$\|v\|_{V(kT,t)} \equiv \|v\|_{V(\Omega \times (kT,t))} = \operatorname*{ess\,sup}_{kT \leq t' \leq t} |v(t)|_{L_2(\Omega)} + |\nabla v|_{L_2(\Omega \times (kT,t))}.$$

The constant $\mathcal{A}$ depends on data but is independent of $k \in \mathbb{N}_0$, so we have in fact a global estimate.

We also show an a priori bound for $\|v_t\|_{V(kT,t)}$, $t \in [kT, (k+1)T]$ in terms of $\|v\|_{L_2(kT,(k+1)T;H^2(\Omega))}$ and this norm of v is estimated in Chaps. 4 and 5.

In this chapter we derive an estimate necessary for the existence of global weak solutions. For this we need homogeneous Dirichlet boundary conditions. This is necessary to perform integrations by parts. To this end, we extend functions (1.3) so that

$$\tilde{d}_i|_{S_2(a_i)} = d_i, \quad i = 1, 2, \quad a_1 = -a, \quad a_2 = a. \tag{3.1}$$

© Springer Nature Switzerland AG 2019

J. Renclawowicz, W. M. Zajączkowski, *The Large Flux Problem to the Navier-Stokes Equations*, Advances in Mathematical Fluid Mechanics, https://doi.org/10.1007/978-3-030-32330-1_3

We use the Hopf function η (see [L1])

$$\eta(\sigma; \varepsilon, \varrho) = \begin{cases} 1 & 0 \le \sigma \le \varrho e^{-1/\varepsilon} \equiv r, \\ -\varepsilon \ln \frac{\sigma}{\varrho} & r < \sigma \le \varrho, \\ 0 & \varrho < \sigma < \infty. \end{cases} \qquad (3.2)$$

We calculate

$$\frac{d\eta}{d\sigma} = \eta'(\sigma; \varepsilon, \varrho) = \begin{cases} 0 & 0 < \sigma \le r, \\ -\frac{\varepsilon}{\sigma} & r < \sigma \le \varrho, \\ 0 & \varrho < \sigma < \infty, \end{cases}$$

so that $|\eta'(\sigma; \varepsilon, \varrho)| \le \frac{\varepsilon}{\sigma}$. We define locally functions η_i in an internal neighborhood of S_2 by setting

$$\eta_i = \eta(\sigma_i; \varepsilon, \varrho), \quad i = 1, 2,$$

where σ_i denotes a local coordinate defined on a small neighborhood of

$$S_2(a_1, \varrho) = \{x \in \Omega : x_3 \in (-a, -a + \varrho)\} \\ S_2(a_2, \varrho) = \{x \in \Omega : x_3 \in (a - \varrho, a)\}, \qquad (3.3)$$

$\sigma_1 = -a + x_3$, $x_3 \in (a, a + \varrho)$ and $\sigma_2 = a - x_3$, $x_3 \in (a - \varrho, a)$. Hence, σ_i, $i = 1, 2$, are positive.

Next we set

$$\alpha = \sum_{i=1}^{2} \tilde{d}_i \eta_i, \quad b = \alpha \bar{e}_3, \quad \bar{e}_3 = (0, 0, 1). \qquad (3.4)$$

Then we introduce the function

$$u = v - b. \qquad (3.5)$$

Therefore

$$\operatorname{div} u = -\operatorname{div} b = -\alpha_{,x_3} \quad \text{in } \Omega, \quad u \cdot \bar{n}|_S = 0.$$

Thus, the boundary conditions for u are homogeneous but the function u is not ideal as the new variable: it is not divergence free. Let us rewrite the compatibility condition as follows

$$\int_\Omega \alpha_{,x_3} dx = -\int_{S_2(-a)} \alpha|_{x_3=-a} dS_2 + \int_{S_2(a)} \alpha|_{x_3=a} dS_2 = 0.$$

We need to correct the function u, so we define φ as a solution to the Neumann problem for the Poisson equation

$$\Delta\varphi = -\operatorname{div} b \qquad \text{in } \Omega,$$
$$\bar{n} \cdot \nabla\varphi = 0 \qquad \text{on } S, \tag{3.6}$$
$$\int_\Omega \varphi \, dx = 0.$$

Next, we set

$$w = u - \nabla\varphi = v - (b + \nabla\varphi) \equiv v - \delta. \tag{3.7}$$

Consequently, (w, p) is a solution to the problem

$$
\begin{aligned}
& w_t + w \cdot \nabla w + w \cdot \nabla\delta + \delta \cdot \nabla w - \operatorname{div} \mathbb{T}(w, p) \\
& \quad = f - \delta_t - \delta \cdot \nabla\delta + \nu \operatorname{div} \mathbb{D}(\delta) \equiv F(\delta, t) && \text{in } \Omega^T, \\
& \operatorname{div} w = 0 && \text{in } \Omega^T, \\
& w \cdot \bar{n} = 0 && \text{on } S^T, \\
& \nu\bar{n} \cdot \mathbb{D}(w) \cdot \bar{\tau}_\alpha + \gamma w \cdot \bar{\tau}_\alpha = -\nu\bar{n} \cdot \mathbb{D}(\delta) \cdot \bar{\tau}_\alpha - \gamma\delta \cdot \bar{\tau}_\alpha \\
& \quad \equiv B_{1\alpha}(\delta), \quad \alpha = 1, 2, && \text{on } S_1^T, \\
& \bar{n} \cdot \mathbb{D}(w) \cdot \bar{\tau}_\alpha = -\bar{n} \cdot \mathbb{D}(\delta) \cdot \bar{\tau}_\alpha \equiv B_{2\alpha}(\delta), \quad \alpha = 1, 2, && \text{on } S_2^T, \\
& w|_{t=0} = v(0) - \delta(0) \equiv w(0) && \text{in } \Omega,
\end{aligned}
\tag{3.8}
$$

where we used that $\operatorname{div}\delta = 0$. Moreover, we set

$$\bar{n}|_{S_1} = \frac{(\varphi_{0,x_1}, \varphi_{0,x_2}, 0)}{\sqrt{\varphi_{0,x_1}^2 + \varphi_{0,x_2}^2}}, \qquad \bar{\tau}_1|_{S_1} = \frac{(-\varphi_{0,x_2}, \varphi_{0,x_1}, 0)}{\sqrt{\varphi_{0,x_1}^2 + \varphi_{0,x_2}^2}},$$
$$\bar{\tau}_2|_{S_1} = (0, 0, 1) = \bar{e}_3, \qquad \bar{n}|_{S_2(-a)} = -\bar{e}_3,$$
$$\bar{n}|_{S_2(a)} = \bar{e}_3, \qquad \bar{\tau}_1|_{S_2} = \bar{e}_1, \qquad \bar{\tau}_2|_{S_2} = \bar{e}_2,$$

where $\bar{e}_1 = (1, 0, 0)$, $\bar{e}_2 = (0, 1, 0)$.

Since Dirichlet boundary conditions for w are homogeneous and w is divergence free, we can consider weak solutions to the problem (3.8).

3.1   Weak Solutions

Definition 3.1 We call w a weak solution to problem (3.8) if for any sufficiently smooth function ψ such that

$$\operatorname{div} \psi|_\Omega = 0, \quad \psi \cdot \bar{n}|_S = 0$$

the integral identity

$$\int_{\Omega^T} w_t \cdot \psi dx dt + \int_{\Omega^T} H(w) \cdot \psi dx dt + \nu \int_{\Omega^T} \mathbb{D}(v) \cdot \mathbb{D}(\psi) dx dt$$

$$+ \gamma \sum_{\alpha=1}^{2} \int_{S_1^T} w \cdot \bar{\tau}_\alpha \psi \cdot \bar{\tau}_\alpha dS_1 dt - \sum_{\alpha,\sigma=1}^{2} \int_{S_\sigma^T} B_{\sigma\alpha} \psi \cdot \bar{\tau}_\alpha dS_\sigma dt = \int_{\Omega^T} F \cdot \psi dx dt$$

holds, where

$$H(w) = w \cdot \nabla w + w \cdot \nabla \delta + \delta \cdot \nabla w.$$

Now, we are able to derive an estimate for the weak solutions to (3.8): in order to obtain the energy estimate we use $\psi = w$ as a test function. Thus, we multiply the first equation in (3.8) by ψ, integrate by parts on Ω, and apply the definition of F. Then we have

$$\frac{1}{2}\frac{d}{dt}|w|_{2,\Omega}^2 + \int_\Omega (w \cdot \nabla \delta \cdot w + \delta \cdot \nabla w \cdot w) dx - \int_\Omega \operatorname{div} \mathbb{T}(w + \delta, p) \cdot w dx$$

$$= \int_\Omega (f - \delta_t - \delta \cdot \nabla \delta) \cdot w dx. \tag{3.9}$$

In view of boundary conditions $(3.8)_{4,5}$ the third integral on the l.h.s. of (3.9) is reformulated as follows

$$-\int_\Omega \operatorname{div} \mathbb{T}(w + \delta, p) \cdot w dx = -\int_\Omega \operatorname{div}\left[\nu \mathbb{D}(w + \delta) - p\mathbb{I}\right] \cdot w dx$$

$$= -\int_\Omega \operatorname{div}\left[\nu \mathbb{D}(w + \delta)\right] \cdot w dx + \int_\Omega \nabla p \cdot w dx$$

$$= \nu \int_\Omega D_{ij}(w + \delta) w_{j,x_i} dx - \nu \int_\Omega \partial_{x_j}[D_{ij}(w + \delta) w_i] dx + \int_\Omega \operatorname{div}(pw) dx$$

$$\equiv I.$$

The first term in I equals

$$\frac{\nu}{2}|D_{ij}(w)|^2_{2,\Omega} + \nu \int_\Omega D_{ij}(\delta)w_{j,x_i}dx,$$

where the summation over repeated indices is assumed. By the Green theorem the second term in I takes the form

$$-\nu \int_{S_1} n_j D_{ij}(w+\delta)w_i dS_1 - \nu \int_{S_2} n_j D_{ij}(w+\delta)w_i dS_2$$

$$= -\nu \int_{S_1} n_j D_{ij}(w+\delta)(w_{\tau_\alpha}\tau_{\alpha i} + w_n n_i)dS_1$$

$$-\nu \int_{S_2} n_j D_{ij}(w+\delta)(w_{\tau_\alpha}\tau_{\alpha i} + w_n n_i)dS_2$$

$$= \gamma \int_{S_1} (|w_{\tau_\alpha}|^2 + w_{\tau_\alpha}\delta_{\tau_\alpha})dS_1,$$

where $w_{\tau_\alpha} = w \cdot \bar{\tau}_\alpha$, $\alpha = 1,2$, $w_n = w \cdot \bar{n}$.

Finally the last term in I vanishes in view of the Green theorem and $(3.8)_3$.

Applying the Korn inequality we derive the inequality

$$\frac{1}{2}\frac{d}{dt}|w|^2_{2,\Omega} + \nu\|w\|^2_{1,\Omega} + \gamma\sum_{\alpha=1}^{2}|w\cdot\bar{\tau}_\alpha|^2_{2,S_1}$$

$$\leq -\int_\Omega (w\cdot\nabla\delta\cdot w + \delta\cdot\nabla w\cdot w)dx + c\sum_{\alpha=1}^{2}|\delta\cdot\bar{\tau}_\alpha|^2_{2,S_1} \qquad (3.10)$$

$$+ c|\mathbb{D}(\delta)|^2_{2,\Omega} + \int_\Omega (f - \delta_{,t} - \delta\cdot\nabla\delta)\cdot w dx.$$

The most difficult terms are those caused by nonlinearity $w \cdot \nabla w$, so let us focus on the integral

$$\int_\Omega \delta\cdot\nabla w\cdot w dx = \int_\Omega (b+\nabla\varphi)\cdot\nabla w\cdot w dx$$

$$= \int_\Omega b\cdot\nabla w\cdot w dx + \int_\Omega \nabla\varphi\cdot\nabla w\cdot w dx \equiv I_1 + I_2. \qquad (3.11)$$

By the Hölder inequality and definition of b, we estimate I_1 by

$$|I_1| \leq |\nabla w|_{2,\Omega}|w|_{6,\Omega}|b|_{3,\Omega} \leq c\|w\|^2_{1,\Omega}|b|_{3,\tilde{S}_2(\varrho)} \leq c\varrho^{1/6}\|w\|^2_{1,\Omega}|b|_{6,\tilde{S}_2(\varrho)}$$

$$\leq c\varrho^{1/6}\|w\|^2_{1,\Omega}\|\tilde{d}\|_{1,\Omega},$$

where $\tilde{S}_2(\varrho) = S_2(a_1, \varrho) \cup S_2(a_2, \varrho)$. Next, we estimate I_2 in the following way

$$|I_2| \leq |\nabla\varphi|_{3,\Omega} |w|_{6,\Omega} |\nabla w|_{2,\Omega},$$

where

$$|\nabla\varphi|_{3,\Omega} \leq c\|\nabla\varphi\|_{L_{3,-\mu'}(\Omega)} \leq c\|\nabla_{x_3}\nabla\varphi\|_{L_{3,1-\mu'}(\Omega)} \leq c\|\varphi\|_{L^2_{3,1-\mu'}(\Omega)}$$
$$\leq c\|\operatorname{div} b\|_{L_{3,1-\mu'}(\Omega)}, \tag{3.12}$$

where $\mu' > 0$ and the results of Chaps. 10 and 11 (see also [RZ2]) are used. Let

$$\tilde{S}'_2(a_1, \varrho) = \{x \in \Omega : x_3 \in (-a + r, -a + \varrho)\},$$
$$\tilde{S}'_2(a_2, \varrho) = \{x \in \Omega : x_3 \in (a - \varrho, a - r)\}.$$

Employing the definition of b and the properties of function η we have for $\mu = 1 - \mu'$ the estimates

$$\|\operatorname{div} b\|_{L_{3,\mu}(\Omega)} \leq c\varepsilon \left(\sum_{i=1}^2 \int_{\tilde{S}'_2(a_i,\varrho)} |\tilde{d}_i|^3 \frac{\sigma_i^{3\mu}}{\sigma_i^3} dx \right)^{1/3}$$
$$+ \left(\sum_{i=1}^2 \int_{\tilde{S}_2(a_i,\varrho)} |\tilde{d}_{i,x_3}|^3 |\sigma_i(x)|^{3\mu} dx \right)^{1/3}$$
$$\leq c \sum_{i=1}^2 \varepsilon \left(\sup_{x_3} \int_{S_2(a_i,\varrho)} |\tilde{d}_i|^3 dx' \int_r^\varrho \frac{\sigma_i^{2\mu}}{\sigma_i^3} d\sigma_i \right)^{1/3} \tag{3.13}$$
$$+ \sum_{i=1}^2 \left(\sup_{x_3} \int_{S_2(a_i,\varrho)} |\tilde{d}_{i,x_3}|^3 dx' \int_0^\varrho \sigma_i^{3\mu} d\sigma_i \right)^{1/3}$$
$$\leq c\varepsilon\varrho^{\mu-2/3} \sup_{x_3} |\tilde{d}|_{3,S_2} + c\varrho^{\mu+1/3} \sup_{x_3} |\tilde{d}_{,x_3}|_{3,S_2},$$

where $\sigma_i = \operatorname{dist}\{S_2(a_i), x\}$, $x \in \tilde{S}_2(a_i, \varrho)$.

The above inequality holds for $\mu > \frac{2}{3}$ because for $\mu = \frac{2}{3}$ the first integral on the r.h.s. of the above inequalities is not integrable.

In view of the definition of the weighted spaces used in (3.12), the weight in (3.12) is $x_3^{2\mu}$, $3\mu > 2$. Therefore, the weight is not the Muckenhoupt weight, so estimate in (3.12) of the singular operator is proved in Chaps. 10 and 11 (see also [RZ1, RZ2]).

Then

$$|I_2| \le c[\varepsilon \varrho^{\mu-2/3} \sup_{x_3} |\tilde{d}|_{3,S_2} + \varrho^{\mu+1/3} \sup_{x_3} |\tilde{d}_{,x_3}|_{3,S_2}] \|w\|_{1,\Omega}^2 \equiv E^2.$$

Now, we consider the term

$$\int_\Omega w \cdot \nabla\delta \cdot w dx = \int_\Omega w \cdot \nabla b \cdot w dx + \int_\Omega w \cdot \nabla\nabla\varphi \cdot w dx \equiv I_3 + I_4. \quad (3.14)$$

For I_4, we have

$$|I_4| = \left| \int_\Omega \text{div}\,(w \cdot \nabla\varphi w) dx - \int_\Omega w \cdot \nabla w \cdot \nabla\varphi dx \right| = \left| \int_\Omega w \cdot \nabla w \cdot \nabla\varphi dx \right| \le E^2$$

because the second term in I_4 is equal to I_2 and the first vanishes in view of $(3.8)_3$.

On the other hand, using $b = \alpha \bar{e}_3 = \sum_{i=1}^2 \tilde{d}_i \eta_i \bar{e}_3$, we find a bound for I_3:

$$|I_3| = \left| \sum_{i=1}^2 \int_{\tilde{S}_2(a_i,\varrho)} w \cdot \nabla(\tilde{d}_i \eta_i) w_3 dx \right|$$

$$= \left| \sum_{i=1}^2 \int_{\tilde{S}_2(a_i,\varrho)} (w \cdot \nabla\tilde{d}_i \eta_i w_3 + w \cdot \nabla\eta_i \tilde{d}_i w_3) dx \right|$$

$$\le \sum_{i=1}^2 \left(\int_{\tilde{S}_2(a_i,\varrho)} |w \cdot \nabla\tilde{d}_i \eta_i|\,|w_3| dx \right.$$

$$\left. + \varepsilon \int_{\tilde{S}_2(a_i,\varrho)} \left| \frac{w_3}{\sigma_i} w_3 \tilde{d}_i \right| \text{supp}\,\eta'_i d\sigma_i dx_1 dx_2 \right)$$

$$\le c \sum_{i=1}^2 |w|_{6,\tilde{S}_2(a_i,\varrho)} |w_3|_{3,\tilde{S}_2(a_i,\varrho)} |\nabla\tilde{d}_i|_{2,\tilde{S}_2(a_i,\varrho)} \quad (3.15)$$

$$+ c\varepsilon \sum_{i=1}^2 |w_3|_{6,\tilde{S}_2(a_i,\varrho)} |\tilde{d}_i|_{3,\tilde{S}_2(a_i,\varrho)} \left(\int_{S_2(a_i)} dx_1 dx_2 \int_r^\varrho d\sigma_i \left| \frac{w_3}{\sigma_i} \right|^2 \right)^{1/2}$$

$$\le c\varrho^{1/6} \sum_{i=1}^2 |w|_{6,\tilde{S}_2(a_i,\varrho)}^2 |\nabla\tilde{d}_i|_{2,\tilde{S}_2(a_i,\varrho)}$$

$$+ c\varepsilon \sum_{i=1}^2 |w|_{6,\tilde{S}_2(a_i,\varrho)} |\nabla w_3|_{2,\tilde{S}_2(a_i,\varrho)} |\tilde{d}_i|_{3,\tilde{S}_2(a_i,\varrho)}$$

$$\le c(\varrho^{1/6} + \varepsilon) \|w\|_{1,\Omega}^2 \|\tilde{d}\|_{1,3,\Omega}.$$

Thus, we can summarize the estimates for I_1–I_4 to conclude that the nonlinear term in (3.10) is bounded by

$$
\left| \int_\Omega (w \cdot \nabla \delta \cdot w + \delta \cdot \nabla w \cdot w) dx \right| \leq c(\varepsilon \varrho^{\mu-2/3} \sup_{x_3} |\tilde{d}|_{3,S_2}
$$
$$
+ \varrho^{\mu+1/3} \sup_{x_3} |\tilde{d}_{,x_3}|_{3,S_2} + (\varrho^{1/6} + \varepsilon)\|\tilde{d}\|_{1,3,\Omega} + \varrho^{1/6}\|\tilde{d}\|_{1,\Omega})\|w\|_{1,\Omega}^2. \tag{3.16}
$$

Next, we examine the second term on the r.h.s. of (3.10),

$$
\sum_{\alpha=1}^2 |\delta \cdot \bar{\tau}_\alpha|_{2,S_1}^2 \leq \sum_{\alpha=1}^2 (|b \cdot \bar{\tau}_\alpha|_{2,S_1}^2 + |\nabla\varphi \cdot \bar{\tau}_\alpha|_{2,S_1}^2)
$$

$$
\leq |\alpha|_{2,S_1}^2 + c\|\nabla\varphi\|_{1,3/2,\Omega}^2 \leq \sum_{i=1}^2 |d_i|_{2,S_1}^2 + c|\operatorname{div} b|_{3/2,\Omega}^2
$$

$$
\leq c\|\tilde{d}\|_{1,3/2,\Omega}^2 + c\sum_{i=1}^2 |\nabla(\tilde{d}_i\eta_i)|_{3/2,\Omega}^2
$$

$$
\leq c\|\tilde{d}\|_{1,3/2,\Omega}^2 + c\sum_{i=1}^2 (|\nabla\tilde{d}_i\eta_i|_{3/2,\Omega}^2 + |\tilde{d}_i\nabla\eta_i|_{3/2,\Omega}^2)
$$

$$
\leq c\|\tilde{d}\|_{1,3/2,\Omega}^2 + c\sum_{i=1}^2 |\tilde{d}_i\nabla\eta_i|_{3/2,\Omega}^2.
$$

We estimate the last expression in more detail

$$
\sum_{i=1}^2 |\tilde{d}_i\nabla\eta_i|_{3/2,\Omega}^2 \leq \varepsilon^2 \left[\left(\int_{-a+r}^{-a+\varrho} dx_3 \int_{S_2(a_1)} dx' \left|\frac{\tilde{d}_1}{a+x_3}\right|^{3/2} \right)^{4/3} \right.
$$
$$
\left. + \left(\int_{a-\varrho}^{a-r} dx_3 \int_{S_2(a_2)} dx' \left|\frac{\tilde{d}_2}{a-x_3}\right|^{3/2} \right)^{4/3} \right]
$$

$$
\leq \varepsilon^2 \left[\sup_{x_3} |\tilde{d}_1|_{3/2,S_2(a_1)}^2 \left(\int_{-a+r}^{-a+\varrho} \left|\frac{1}{a+x_3}\right|^{3/2} dx_3 \right)^{4/3} \right.
$$
$$
\left. + \sup_{x_3} |\tilde{d}_2|_{3/2,S_2(a_2)}^2 \left(\int_{a-\varrho}^{a-r} \left|\frac{1}{a-x_3}\right|^{3/2} dx_3 \right)^{4/3} \right]
$$

$$
\leq c\varepsilon^2 \sup_{x_3} |\tilde{d}|_{3/2,S_2}^2 \left(\int_r^\varrho \frac{dy}{y^{3/2}} \right)^{4/3} \leq c\varepsilon^2 \sup_{x_3} |\tilde{d}|_{3/2,S_2}^2 [r^{-1/2} - \varrho^{-1/2}]^{4/3}
$$

$$\leq c\varepsilon^2 \sup_{x_3} |\tilde{d}|^2_{3/2,S_2} \frac{1}{\varrho^{2/3}} \left[\exp\left(\frac{1}{2\varepsilon}\right) - 1\right]^{4/3} \leq c\frac{\varepsilon^2}{\varrho^{2/3}} \exp\left(\frac{2}{3\varepsilon}\right) \sup_{x_3} |\tilde{d}|^2_{3/2,S_2}.$$

Combining the inequalities above, we infer

$$\sum_{\alpha=1}^{2} |\delta \cdot \bar{\tau}_\alpha|^2_{2,S_1} \leq c\|\tilde{d}\|^2_{1,3/2,\Omega} + c\frac{\varepsilon^2}{\varrho^{2/3}} \exp\left(\frac{2}{3\varepsilon}\right) \sup_{x_3} |\tilde{d}|^2_{3/2,S_2}. \qquad (3.17)$$

We also estimate the term

$$|\mathbb{D}(\delta)|^2_{2,\Omega} \leq |\mathbb{D}(b)|^2_{2,\Omega} + |\mathbb{D}(\nabla\varphi)|^2_{2,\Omega}$$

$$\leq \sum_{i=1}^{2} (|\nabla \tilde{d}_i \eta_i|^2_{2,\Omega} + |\tilde{d}_i \nabla \eta_i|^2_{2,\Omega}) + |\nabla^2 \varphi|^2_{2,\Omega}$$

$$\leq c\sum_{i=1}^{2} (|\nabla \tilde{d}_i \eta_i|^2_{2,\Omega} + |\tilde{d}_i \nabla \eta_i|^2_{2,\Omega})$$

$$\leq c\sum_{i=1}^{2} \|\tilde{d}_i\|_{1,2,\Omega} + c\varepsilon^2 \int_{-a+r}^{-a+\varrho} dx_3 \int_{S_2(a_1)} dx' \left|\frac{\tilde{d}_1}{a+x_3}\right|^2$$

$$+ \varepsilon^2 c \int_{a-\varrho}^{a-r} dx_3 \int_{S_2(a_2)} dx' \left|\frac{\tilde{d}_2}{a-x_3}\right|^2 \qquad (3.18)$$

$$\leq c\sum_{i=1}^{2} \left(\|\tilde{d}_i\|^2_{1,2,\Omega} + \varepsilon^2 \sup_{x_3} |\tilde{d}_i|^2_{2,S_2} \int_r^\varrho \frac{dy}{y^2}\right)$$

$$\leq c\sum_{i=1}^{2} \left[\|\tilde{d}_i\|^2_{1,2,\Omega} + \varepsilon^2 \sup_{x_3} |\tilde{d}_i|^2_{2,S_2} \left(\frac{1}{r} - \frac{1}{\varrho}\right)\right]$$

$$\leq c\sum_{i=1}^{2} \left[\|\tilde{d}_i\|^2_{1,2,\Omega} + \varepsilon^2 \sup_{x_3} |\tilde{d}_i|^2_{2,S_2} \frac{1}{\varrho} \left(\exp\left(\frac{1}{\varepsilon}\right) - 1\right)\right]$$

$$\leq c\sum_{i=1}^{2} [\|\tilde{d}_i\|^2_{1,2,\Omega} + \frac{\varepsilon^2}{\varrho} e^{1/\varepsilon} \sup_{x_3} |\tilde{d}_i|^2_{2,S_2}].$$

Analyzing the last integral on the r.h.s. of (3.10) we have

$$\int_\Omega (f - \delta_t - \delta \cdot \nabla\delta) \cdot w\, dx \leq \varepsilon_1 |w|^2_{6,\Omega} + c(1/\varepsilon_1)(|f|^2_{6/5,\Omega} + |\delta_t|^2_{6/5,\Omega}) + \left|\int_\Omega \delta \cdot \nabla\delta \cdot w\, dx\right|.$$

We estimate

$$
\begin{aligned}
|\delta_t|_{6/5,\Omega} = |b_t + \nabla\varphi_t|_{6/5,\Omega} &\leq |\tilde{d}_t|_{6/5,\Omega} + |\mathrm{div}\, b_t|_{6/5,\Omega} \\
&\leq |\tilde{d}_t|_{6/5,\Omega} + |\nabla\tilde{d}_t|_{6/5,\Omega} + |\tilde{d}_t\nabla\eta|_{6/5,\Omega} \leq \|\tilde{d}_t\|_{1,6/5,\Omega} \\
&\quad + \varepsilon\sup_{x_3}|\tilde{d}_t|_{6/5,S_2}\left(\int_r^\varrho \frac{dx_3}{x_3^{6/5}}\right)^{5/6} \\
&\leq \|\tilde{d}_t\|_{1,6/5,\Omega} + c\frac{\varepsilon}{\varrho^{1/6}}e^{1/6\varepsilon}\sup_{x_3}|\tilde{d}_t|_{6/5,S_2},
\end{aligned}
\tag{3.19}
$$

where we used that

$$
\left(\int_r^\varrho \frac{dx_3}{x_3^{6/5}}\right)^{5/6} = \left(5\frac{1}{x_3^{1/5}}\Big|_\varrho^r\right)^{5/6} = 5^{5/6}\left(\frac{1}{r^{1/5}} - \frac{1}{\varrho^{1/5}}\right)^{5/6} = \frac{5^{5/6}}{\varrho^{1/6}}(e^{1/5\varepsilon}-1)^{5/6}.
$$

Finally, we examine

$$
\left|\int_\Omega \delta\cdot\nabla\delta\cdot w\,dx\right| \leq |\nabla\delta|_{2,\Omega}|\delta|_{3,\Omega}|w|_{6,\Omega} \leq \varepsilon_2|w|_{6,\Omega}^2 + c(1/\varepsilon_2)\|\delta\|_{1,2,\Omega}^4
\tag{3.20}
$$

$$
\leq \varepsilon_2|w|_{6,\Omega}^2 + c(1/\varepsilon_2)\left(\|\tilde{d}\|_{1,2,\Omega}^4 + \frac{\varepsilon^4}{\varrho^2}e^{2/\varepsilon}\sup_{x_3}|\tilde{d}|_{2,\Omega}^4\right).
$$

Employing the above estimates in (3.10) yields

$$
\begin{aligned}
\frac{1}{2}\frac{d}{dt}|w|_{2,\Omega}^2 + \nu\|w\|_{1,\Omega}^2 &+ \gamma\sum_{\alpha=1}^2|w\cdot\bar\tau_\alpha|_{2,S_1}^2 \\
&\leq c\|w\|_{1,\Omega}^2[\varepsilon\varrho^{\mu-2/3}\sup_{x_3}|\tilde{d}|_{3,S_2} + \varrho^{\mu+1/3}\sup_{x_3}|\tilde{d},_{x_3}|_{3,S_2} \\
&\quad + (\varrho^{1/6} + \varepsilon)\|\tilde{d}\|_{1,3,\Omega}] \\
&\quad + c[|f|_{6/5,\Omega}^2 + \|\tilde{d}\|_{1,\Omega}^4 + \|\tilde{d}\|_{1,\Omega}^2 + \|\tilde{d}_t\|_{1,6/5,\Omega}^2 \\
&\quad + \frac{\varepsilon^2}{\varrho}e^{1/\varepsilon}\sup_{x_3}|\tilde{d}|_{2,S_2}^2 + \frac{\varepsilon^4}{\varrho^2}e^{2/\varepsilon}\sup_{x_3}|\tilde{d}|_{2,S_2}^4 + \frac{\varepsilon^2}{\varrho^{2/3}}e^{2/3\varepsilon}\sup_{x_3}|\tilde{d}|_{3/2,S_2}^2 \\
&\quad + \frac{\varepsilon^2}{\varrho^{1/3}}e^{1/3\varepsilon}\sup_t|\tilde{d}_t|_{6/5,S_2}^2].
\end{aligned}
\tag{3.21}
$$

Introduce the anisotropic Sobolev spaces

$$
\|u\|_{W_{p_1,p_2}^1(\Omega)} = \left(\int_{-a}^a\left(\int_{S_2}\sum_{|\alpha|\leq 1}|D_x^\alpha u|^{p_1}\,dx_1dx_2\right)^{p_2/p_1}dx_3\right)^{1/p_2},
$$

$$\|u\|_{L_{p_1,p_2}(\Omega)} = \left[\left(\int_{S_2} |u|^{p_1} dx_1 dx_2\right)^{p_2/p_1} dx_3\right]^{1/p_2},$$

where $p_1, p_2 \in [1, \infty]$. Therefore, the norms of $\tilde{d}$ under the first square bracket on the r.h.s. of (3.21) are bounded by

$$\|\tilde{d}\|_{W^1_{3,\infty}(\Omega)},$$

and the first three norms of $\tilde{d}$ under the second square bracket are bounded by the quantities

$$\|\tilde{d}\|_{1,\Omega}, \quad \|\tilde{d}_t\|_{1,6/5,\Omega}.$$

Then, we express (3.21) in the form

$$\frac{1}{2}\frac{d}{dt}|w|^2_{2,\Omega} + \nu\|w\|^2_{1,\Omega} + \gamma\sum_{\alpha=1}^{2}|w\cdot\bar{\tau}_\alpha|^2_{2,S_1}$$

$$\leq c\|w\|^2_{1,\Omega}[(\varepsilon\varrho^{\mu-2/3} + \varrho^{\mu+1/3} + \varrho^{1/6} + \varepsilon)\|\tilde{d}\|_{W^1_{3,\infty}(\Omega)}]$$

$$+ c\Big[|f|^2_{6/5,\Omega} + \|\tilde{d}\|^2_{1,\Omega} + \|\tilde{d}\|^4_{1,\Omega} + \|\tilde{d}_t\|^2_{1,6/5,\Omega} \tag{3.22}$$

$$+ \left(\frac{\varepsilon^2}{\varrho}e^{1/\varepsilon} + \frac{\varepsilon^2}{\varrho^{2/3}}e^{2/3\varepsilon}\right)\|\tilde{d}\|^2_{L_{2,\infty}(\Omega)} + \frac{\varepsilon^4}{\varrho^2}e^{2/\varepsilon}\|\tilde{d}\|^4_{L_{2,\infty}(\Omega)}$$

$$+ \frac{\varepsilon^2}{\varrho^{1/3}}e^{1/3\varepsilon}\|\tilde{d}_t\|^2_{L_{6/5,\infty}(\Omega)}\Big].$$

We set $\mu > \frac{2}{3}$ and $\varrho < 1$. Then $\varrho^{\mu+1/3} < \varrho^{1/6}$. Therefore, the expression under the first square bracket in (3.22) is bounded by

$$(\varepsilon + \varrho^{1/6})\|\tilde{d}\|_{W^1_{3,\infty}(\Omega)}.$$

Setting

$$c(\varepsilon + \varrho^{1/6})\|\tilde{d}\|_{W^1_{3,\infty}(\Omega)} \leq \frac{\nu}{2} \tag{3.23}$$

we derive from (3.22) the inequality

$$\frac{d}{dt}|w|_{2,\Omega}^2 + \nu\|w\|_{1,\Omega}^2 + 2\gamma\sum_{\alpha=1}^{2}|w\cdot\bar{\tau}_\alpha|_{2,S_1}^2$$

$$\leq c[|f|_{6/5,\Omega}^2 + \|\tilde{d}\|_{1,\Omega}^2 + \|\tilde{d}\|_{1,\Omega}^4 + \|\tilde{d}_t\|_{1,6/5,\Omega}^2]$$

$$+ c\left[\left(\frac{\varepsilon^2}{\varrho}e^{1/\varepsilon} + \frac{\varepsilon^2}{\varrho^{2/3}}e^{2/3\varepsilon}\right)\|\tilde{d}\|_{L_{2,\infty}(\Omega)}^2\right. \tag{3.24}$$

$$\left.+ \frac{\varepsilon^4}{\varrho^2}e^{2/\varepsilon}\|\tilde{d}\|_{L_{2,\infty}(\Omega)}^4 + \frac{\varepsilon^2}{\varrho^{1/3}}e^{1/3\varepsilon}\|\tilde{d}_t\|_{L_{6/5,\infty,\Omega}}^2\right].$$

From (3.23) we calculate

$$\varepsilon = \frac{\nu}{4c\|\tilde{d}\|_{W_{3,\infty}^1(\Omega)}}, \qquad \varrho = \left(\frac{\nu}{4c\|\tilde{d}\|_{W_{3,\infty}^1(\Omega)}}\right)^6. \tag{3.25}$$

Then (3.24) takes the form

$$\frac{d}{dt}|w|_{2,\Omega}^2 + \nu\|w\|_{1,\Omega}^2 + 2\gamma\sum_{\alpha=1}^{2}|w\cdot\bar{\tau}_\alpha|_{2,S_1}^2$$

$$\leq c[|f|_{6/5,\Omega}^2 + \|\tilde{d}\|_{1,\Omega}^2 + \|\tilde{d}\|_{1,\Omega}^4 + \|\tilde{d}_t\|_{1,6/5,\Omega}^2]$$

$$+ c\left[(1 + \|\tilde{d}\|_{W_{3,\infty}^1(\Omega)}^2)\|\tilde{d}\|_{W_{3,\infty}^1(\Omega)}^4 \exp\left(\frac{1}{c}\|\tilde{d}\|_{W_{3,\infty}^1(\Omega)}^2\right)\right. \tag{3.26}$$

$$\left.+ \|\tilde{d}\|_{W_{3,\infty}^1(\Omega)}^{10}\exp\left(\frac{1}{c}\|\tilde{d}\|_{W_{3,\infty}^1(\Omega)}\right) + \exp\left(\frac{1}{c}\|\tilde{d}\|_{W_{3,\infty}^1(\Omega)}\right)\|\tilde{d}_t\|_{L_{6/5,\infty,\Omega}}^2\right],$$

where we used the following calculations. Let $a = \|\tilde{d}\|_{W_{3,\infty}^1(\Omega)}$, $\varepsilon = \frac{c}{a}$, $\rho = \left(\frac{c}{a}\right)^6$. Then the first term under the second square bracket in the r.h.s. of (3.24) equals

$$\left[\frac{\left(\frac{c}{a}\right)^2}{\left(\frac{c}{a}\right)^6}e^{a/c} + \frac{\left(\frac{c}{a}\right)^2}{\left(\frac{c}{a}\right)^4}e^{\frac{2}{3}\frac{a}{c}}\right]a^2 \leq c(a^2 + 1)a^4 e^{a/c},$$

the second

$$\frac{\left(\frac{c}{a}\right)^4}{\left(\frac{c}{a}\right)^{12}}e^{2a/c}a^2 = ca^8 e^{2a/c}a^2,$$

and the third

$$\frac{\left(\frac{c}{a}\right)^2}{\left(\frac{c}{a}\right)^2} e^{a/3c} \|\tilde{d}_t\|^2_{L_{6/5,\infty},\Omega} \le c e^{a/c} \|\tilde{d}_t\|^2_{L_{6/5,\infty},\Omega}.$$

To simplify notation we write (3.26) in the form

$$\frac{d}{dt}|w|^2_{2,\Omega} + \nu\|w\|^2_{1,\Omega} \le c[|f|^2_{6/5,\Omega} + \varphi(\|\tilde{d}\|_{W^1_{3,\infty}(\Omega)}) \cdot$$
$$\cdot (\|\tilde{d}\|^2_{W^1_{3,\infty}(\Omega)} + \|\tilde{d}_t\|^2_{1,6/5,\Omega})] \equiv F^2(t), \tag{3.27}$$

where φ is an increasing positive function which form follows from the form of the r.h.s. of (3.26) and the following estimates

$$\|u\|_{L_{6/5,\infty}(\Omega)} \le c\|u\|_{1,6/5,\Omega},$$
$$\|u\|_{1,\Omega} \le c\|u\|_{W^1_{3,\infty}(\Omega)}.$$

To obtain a global estimate we write (3.27) in the form

$$\frac{d}{dt}(|w|^2_{2,\Omega}e^{\nu t}) \le F^2(t)e^{\nu t}. \tag{3.28}$$

Integrating (3.28) with respect to time from $t = kT$, to $t \in (kT, (k+1)T]$, $k \in \mathbb{N}_0$, we obtain

$$|w(t)|^2_{2,\Omega} \le e^{-\nu t}\int_{kT}^t F^2(t')e^{\nu t'}\,dt' + \exp(-\nu(t - kT))|w(kT)|^2_{2,\Omega}. \tag{3.29}$$

Setting $t = (k+1)T$ we get

$$|w((k+1)T)|^2_{2,\Omega} \le \int_{kT}^{(k+1)T} F^2(t)dt + \exp(-\nu T)|w(kT)|^2_{2,\Omega}. \tag{3.30}$$

Let

$$A_1^2(T) = \sup_{k \in \mathbb{N}_0} \int_{kT}^{(k+1)T} F^2(t)dt. \tag{3.31}$$

Then (3.30), by iteration, implies the estimate

$$|w(kT)|^2_{2,\Omega} \le \frac{A_1^2(T)}{1 - \exp(-\nu T)} + \exp(-\nu kT)|w(0)|^2_{2,\Omega}. \tag{3.32}$$

Next, we integrate (3.27) with respect to time from $t = kT$ to $t \in (kT,(k+1)T]$. Using (3.32) yields

$$
\begin{aligned}
\|w\|^2_{V(\Omega \times (kT,t))} &\leq \int_{kT}^{t} F^2(t')dt' + |w(kT)|^2_{2,\Omega} \\
&\leq A_1^2(T) + \frac{A_1^2(T)}{1 - \exp(-\nu T)} + \exp(-\nu kT)|w(0)|^2_{2,\Omega} \\
&= A_1^2(T)\frac{2 - \exp(-\nu T)}{1 - \exp(-\nu T)} + \exp(-\nu kT)|w(0)|^2_{2,\Omega} \leq A_2^2(T),
\end{aligned}
\tag{3.33}
$$

where

$$
A_2^2(T) = A_1^2(T)\frac{2 - \exp(-\nu T)}{1 - \exp(-\nu T)} + |w(0)|^2_{2,\Omega}.
\tag{3.34}
$$

Having estimate (3.33) for w and using transformation (3.7) we derive

$$
\|v\|^2_{V(\Omega \times (kT,t))} \leq \|w\|^2_{V(\Omega \times (kT,t))} + \|\delta\|^2_{V(\Omega \times (kT,t))}.
\tag{3.35}
$$

We restrict our considerations to the time interval $(0,T)$ only because estimates in any interval $[kT,(k+1)T]$ can be performed in the same way.

The last term in the r.h.s. of (3.35) is bounded by

$$
\begin{aligned}
\|\delta\|^2_{V(\Omega^t)} &\leq \|b\|^2_{V(\Omega^t)} + \|\nabla\varphi\|^2_{V(\Omega^t)} \leq |b|^2_{2,\infty,\Omega^t} + \|b\|^2_{1,2,2,\Omega^t} \\
&\quad + |\nabla\varphi|^2_{2,\infty,\Omega^t} + \|\nabla\varphi\|^2_{1,2,2,\Omega^t} \equiv I.
\end{aligned}
\tag{3.36}
$$

Now, we estimate the particular terms in I.

The first term

$$
|b|^2_{2,\infty,\Omega^t} \leq |\alpha|^2_{2,\infty,\Omega^t} \leq \sum_{i=1}^{2} |\tilde{d}_i|^2_{2,\infty,\Omega^t} = |\tilde{d}|_{2,\infty,\Omega^t}.
$$

The second term

$$
\begin{aligned}
\|b\|^2_{1,2,2,\Omega^t} &= \int_0^t (|\nabla b|^2_{2,\Omega} + |b|^2_{2,\Omega})dt' \leq \sum_{i=1}^{2} \int_0^t |\nabla\tilde{d}_i\eta_i|^2_{2,\Omega}dt' \\
&\quad + \sum_{i=1}^{2} \int_0^t |\tilde{d}_i\nabla\eta_i|^2_{2,\Omega}dt' + \sum_{i=1}^{2} \int_0^t |\tilde{d}_i|^2_{2,\Omega}dt'
\end{aligned}
$$

$$\leq c\|\tilde{d}\|_{1,2,2,\Omega^t}^2 + \varepsilon^2 \left[\int_0^t \int_{-a+r}^{-a+\varrho} dx_3 \int_{S_2(a_1)} dx' \left| \frac{\tilde{d}_1}{a + x_3} \right|^2 \right.$$

$$\left. + \int_0^t dt' \int_{a-\varrho}^{a-r} dx_3 \int_{S_2(a_2)} dx' \left| \frac{\tilde{d}_2}{a - x_3} \right|^2 \right]$$

$$\leq c\|\tilde{d}\|_{1,2,2,\Omega^t}^2 + \varepsilon^2 \int_0^t \left[\sup_{x_3} |\tilde{d}_1|_{2,S_2(a_1)}^2 \int_{-a+r}^{-a+\varrho} \left| \frac{1}{a + x_3} \right|^2 dx_3 \right.$$

$$\left. + \sup_{x_3} |\tilde{d}_2|_{2,S_2(a_2)}^2 \int_{a-\varrho}^{a-r} \left| \frac{1}{a - x_3} \right|^2 dx_3 \right] dt'$$

$$\leq c\|\tilde{d}\|_{1,2,2,\Omega^t}^2 + c\varepsilon^2 \int_0^t \sup_{x_3} |\tilde{d}|_{2,S_2}^2 dt' \int_r^\varrho \frac{dy}{y^2}$$

$$= c\|\tilde{d}\|_{1,2,2,\Omega^t}^2 + c\varepsilon^2 \int_0^t \sup_{x_3} |\tilde{d}|_{2,S_2}^2 dt' \left(\frac{1}{r} - \frac{1}{\varrho} \right) =$$

$$= c\|\tilde{d}\|_{1,2,2,\Omega^t}^2 + c\varepsilon^2 \int_0^t \sup_{x_3} |\tilde{d}|_{2,S_2}^2 dt' \frac{1}{\varrho} \left[\exp\left(\frac{1}{\varepsilon} \right) - 1 \right]$$

$$\leq c\|\tilde{d}\|_{1,2,2,\Omega^t} + c\frac{\varepsilon^2}{\varrho} \exp\left(\frac{1}{\varrho} \right) \int_0^t \sup_{x_3} |\tilde{d}|_{2,S_2}^2 dt' \equiv I_1.$$

Employing (3.25) gives

$$I_1 \leq c\|\tilde{d}\|_{1,2,2,\Omega^t}^2 + c\sup_t \|\tilde{d}\|_{W_{3,\infty}^1(\Omega)} \exp\left(\frac{\sup_t \|\tilde{d}\|_{W_{3,\infty}^1(\Omega)}}{c} \right)$$

$$\cdot \int_0^t \sup_{x_3} |\tilde{d}|_{2,S_2}^2 dt' \equiv J. \tag{3.37}$$

Next, we consider the third term in I (see (3.36))

$$|\nabla\varphi|_{2,\infty,\Omega^t}^2 = \sup_t |\nabla\varphi(t')|_{2,\Omega}^2.$$

Multiplying $(3.6)_1$ by φ, integrating by parts, and using $(3.6)_2$ give

$$|\nabla\varphi|_{2,\Omega}^2 = \int_\Omega \operatorname{div} b\, \varphi dx = \int_\Omega \operatorname{div}(b\varphi)dx - \int_\Omega b \cdot \nabla\varphi dx$$

$$= \sum_{i=1}^2 \int_{S_2(a_i)} d_i \varphi dS_2 - \int_\Omega b \cdot \nabla\varphi dx. \tag{3.38}$$

Applying the Hölder and Young inequalities to the r.h.s. of (3.38) and using the Poincaré inequality to the l.h.s. of (3.38) we derive

$$\|\varphi\|_{1,\Omega}^2 \leq \varepsilon_1 |\nabla\varphi|_{2,\Omega}^2 + \frac{c}{\varepsilon_1}|b|_{2,\Omega}^2 + \varepsilon_2|\varphi|_{2,S_2}^2 + \frac{c}{\varepsilon_2}\sum_{i=1}^2 |\tilde{d}_i|_{2,S_2(a_i)}^2.$$

Hence, for sufficiently small ε_1 and ε_2 we have

$$\|\varphi\|_{1,\Omega}^2 \leq c|b|_{2,\Omega}^2 + c|d|_{2,S_2}^2 \leq c|d|_{L_{2,\infty}(\Omega)}^2.$$

Then it follows

$$\|\varphi\|_{1,2,\infty,\Omega^t} \leq c\sup_{t'\leq t}|d(t')|_{L_{2,\infty}(\Omega)}. \tag{3.39}$$

Finally, we estimate the last term in I (see (3.36))

$$\|\nabla\varphi\|_{1,2,2,\Omega^t}^2 = \int_0^t (|\nabla^2\varphi|_{2,\Omega}^2 + |\nabla\varphi|_{2,\Omega}^2)dt' \equiv I_2.$$

In view of solvability of problem (3.6) we have

$$I_2 \leq c\int_0^t |d(t')|_{L_{2,\infty}(\Omega)}^2 dt' + c\int_0^t |\operatorname{div} b(t')|_{2,\Omega}^2 dt' \equiv I_3.$$

The second term in I_3 is bounded by

$$c\int_0^t \sum_{i=1}^2 (|\nabla\tilde{d}_i|_{2,\Omega}^2 + |\tilde{d}_i\nabla\eta_i|_{2,\Omega}^2)dt' \leq cJ,$$

where J is introduced in (3.37). Summarizing, we obtain the estimate

$$\|\delta\|_{V(\Omega^t)}^2 \leq c|\tilde{d}|_{2,\infty,\Omega^t}^2 + c\|\tilde{d}\|_{1,2,2,\Omega^t}^2$$

$$+c\sup_t \|\tilde{d}\|_{W_{3,\infty}^1(\Omega)}^4 \exp\left(\frac{1}{c}\sup_t \|\tilde{d}\|_{W_{3,\infty}^1(\Omega)}\right) \int_0^t \sup_{x_3} |\tilde{d}|_{2,S_2}^2 dt' \tag{3.40}$$

$$+c\sup_{t'\leq t}\|d(t')\|_{L_{2,\infty}(\Omega)}^2.$$

A similar estimate holds for any time interval $[kT, (k+1)T]$, $k \in \mathbb{N}_0$.

The above considerations imply the following result.

Lemma 3.2 *Let (v,p) be a solution to problem (1.1). Let w be defined by (3.8). Then*

(1)

$$\|w\|^2_{V(\Omega \times (kT,t))} \leq cA_1^2(T) + \exp(-\nu kT)|w(0)|^2_{2,\Omega},$$

where $T > 0$, $t \in [kT, (k+1)T]$, $k \in \mathbb{N}_0$, and

$$A_1^2(T) = \sup_{k \in \mathbb{N}_0} \int_{kT}^{(k+1)T} (|f(t)|^2_{6/5,\Omega} + \|d(t)\|^2_{1,3,S_2}$$

$$+\|d_t(t)\|^2_{6/5,S_2})dt \, \varphi(\sup_t \|d(t)\|_{1,3,S_2}) < \infty,$$

where φ is an increasing positive function. Next

(2)

$$\|v\|^2_{V(\Omega \times (kT,t))} \leq \|w\|^2_{V(\Omega \times (kT,t))} + \varphi(\sup_t \|d(t)\|_{1,3,S_2}) \cdot$$

$$\cdot \|d\|^2_{\frac{1}{2},2,S_2 \times (kT,t)} \equiv \|w\|^2_{V(\Omega \times (kT,t))} + F_1^2(t),$$

where $d \in L_\infty(kT, (k+1)T; W_3^1(S_2)) \cap L_2(kT, (k+1)T; H^{1/2}(S_2))$ for any $k \in \mathbb{N}_0$.

Proof Property (1) follows from (3.33) and property (2) from (3.35) and (3.40). This ends the proof.   □

3.2   Estimates for v_t

Consider the problem

$$
\begin{aligned}
v_{tt} - \operatorname{div} \mathbb{T}(v_t, p_t) &= -v \cdot \nabla v_t - v_t \cdot \nabla v + f_t & &\text{in } \Omega^T, \\
\operatorname{div} v_t &= 0 & &\text{in } \Omega^T, \\
v_t \cdot \bar{n} &= 0 & &\text{on } S_1^T, \\
\nu \bar{n} \cdot \mathbb{D}(v_t) \cdot \bar{\tau}_\alpha + \gamma v_t \cdot \bar{\tau}_\alpha &= 0, \quad \alpha = 1,2, & &\text{on } S_1^T, \quad (3.41) \\
v_t \cdot \bar{n} &= d_t & &\text{on } S_2^T, \\
\bar{n} \cdot \mathbb{D}(v_t) \cdot \bar{\tau}_\alpha &= 0, \quad \alpha = 1,2 & &\text{on } S_2^T, \\
v_t|_{t=0} &= v_t(0) & &\text{in } \Omega.
\end{aligned}
$$

Now, instead of (3.4)–(3.7) we have

$$\alpha_t = \sum_{i=1}^{2} \tilde{d}_{i,t} \eta_i, \quad b_t = \alpha_t \bar{e}_3 \tag{3.42}$$

$$u_t = v_t - b_t, \quad \operatorname{div} u_t = -\alpha_{,x_3 t}, \quad u_t \cdot \bar{n}|_S = 0,$$

$$\Delta \varphi_t = -\operatorname{div} b_t \qquad \text{in } \Omega,$$

$$\bar{n} \cdot \nabla \varphi_t = 0 \qquad\qquad \text{on } S, \tag{3.43}$$

$$\int_\Omega \varphi_t dx = 0,$$

$$w_t = u_t - \nabla \varphi_t = v_t - (b_t + \nabla \varphi_t) = v_t - \delta_t. \tag{3.44}$$

Consequently, (w_t, p_t) is a solution to the problem

$$w_{tt} + w \cdot \nabla w_t + w \cdot \nabla \delta_t + \delta \cdot \nabla w_t + w_t \cdot \nabla w + \delta_t \cdot \nabla w$$

$$+ w_t \cdot \nabla \delta - \operatorname{div} \mathbb{T}(w_t, p_t) = -\delta_{tt} - \delta \cdot \nabla \delta_t - \delta_t \cdot \nabla \delta$$

$$+ \nu \operatorname{div} \mathbb{D}(\delta_t) + f_t \equiv F_1(\delta, \delta_t, t) \quad \text{in } \Omega^T,$$

$$\operatorname{div} w_t = 0 \quad \text{in } \Omega^T,$$

$$w_t \cdot \bar{n} = 0 \quad \text{on } S^T, \tag{3.45}$$

$$\nu \bar{n} \cdot \mathbb{D}(w_t) \cdot \bar{\tau}_\alpha + \gamma w_t \cdot \bar{\tau}_\alpha$$

$$= -\nu \bar{n} \cdot \mathbb{D}(\delta_t) \cdot \bar{\tau}_\alpha - \gamma \delta_t \cdot \bar{\tau}_\alpha \quad \alpha = 1, 2 \quad \text{on } S_1^T,$$

$$\bar{n} \cdot \mathbb{D}(w_t) \cdot \bar{\tau}_\alpha = -\bar{n} \cdot \mathbb{D}(\delta_t) \cdot \bar{\tau}_\alpha, \quad \alpha = 1, 2 \quad \text{on } S_2^T,$$

$$w_t|_{t=0} = v_t(0) - \delta_t(0) \equiv w_t(0) \quad \text{in } \Omega.$$

Multiply (3.45) by w_t and integrate over Ω. Then we derive

$$\frac{1}{2} \frac{d}{dt} |w_t|_{2,\Omega}^2 + \int_\Omega (w \cdot \nabla w_t + w_t \cdot \nabla w + w \cdot \nabla \delta_t + \delta \cdot \nabla w_t + \delta_t \cdot \nabla w$$

$$+ w_t \cdot \nabla \delta) \cdot w_t dx - \int_\Omega \operatorname{div} \mathbb{T}(w_t + \delta_t, p_t) \cdot w_t dx \tag{3.46}$$

$$= \int_\Omega (f_t - \delta_{tt} - \delta \cdot \nabla \delta_t - \delta_t \cdot \nabla \delta) \cdot w_t dx.$$

In view of the boundary conditions the third integral on the l.h.s. of (3.46) assumes the form

$$\nu \int_\Omega D_{ij}(w_t + \delta_t)w_{tj,x_i}dx - \nu \int_\Omega \partial_{x_j}[D_{ij}(w_t + \delta_t)w_{ti}] + \int_\Omega \text{div}\,(p_t w_t)dx \equiv I,$$

where the first integral equals

$$\frac{\nu}{2}|D_{ij}(w_t)|^2_{2,\Omega} + \nu \int_\Omega D_{ij}(\delta_t)w_{tj,x_i}dx,$$

the second has the form

$$\gamma \int_{S_1} (|w_{t\tau_\alpha}|^2 + w_{t\tau_\alpha}\delta_{t\tau_\alpha})dS_1,$$

and the third vanishes in view of the Green theorem and $(3.45)_3$. Applying the Korn inequality (see Lemma 2.22) we obtain the inequality

$$\frac{1}{2}\frac{d}{dt}|w_t|^2_{2,\Omega} + \nu\|w_t\|^2_{1,\Omega} + \gamma\sum_{\alpha=1}^{2}|w_t \cdot \bar{\tau}_\alpha|^2_{2,S_1}$$

$$\leq -\int_\Omega (w \cdot \nabla w_t + w_t \cdot \nabla w + w \cdot \nabla\delta_t + \delta \cdot \nabla w_t$$

$$+\delta_t \cdot \nabla w + w_t \cdot \nabla\delta) \cdot w_t dx + c|D(\delta_t)|^2_{2,\Omega} + c\sum_{\alpha=1}^{2}|\delta_t \cdot \bar{\tau}_\alpha|^2_{2,S_1} \tag{3.47}$$

$$+ \int_\Omega (f_t - \delta_{tt} - \delta \cdot \nabla\delta_t - \delta_t \cdot \nabla\delta) \cdot w_t dx.$$

In view of $(3.45)_{2,3}$ the first integral in the first term on the r.h.s. of (3.47) vanishes. The second can be written in the form

$$\int_\Omega w_t \cdot \nabla w_t \cdot w dx$$

and estimated by

$$\varepsilon_1|\nabla w_t|^2_{2,\Omega} + c/\varepsilon_1|w|^2_{\infty,\Omega}|w_t|^2_{2,\Omega}.$$

Now, we shall estimate the other terms from the first integral on the r.h.s. of (3.47). The third term has the form

$$\int_\Omega w \cdot \nabla\delta_t \cdot w_t dx = \int_\Omega w \cdot \nabla b_t \cdot w_t dx + \int_\Omega w \cdot \nabla\nabla\varphi_t w_t dx \equiv J_1 + J_2.$$

For J_2 we have

$$|J_2| = \left| \int_\Omega w \cdot \nabla \varphi_t \nabla w_t dx \right| \leq |\nabla w_t|_{2,\Omega} |w|_{6,\Omega} |\nabla \varphi_t|_{3,\Omega}.$$

In view of (3.12) and (3.13) we obtain the estimate

$$|J_2| \leq c(\varepsilon \varrho^{\mu-2/3} \sup_{x_3} |\tilde{d}_{,t}|_{3,S_2} + \varrho^{\mu+1/3} \sup_{x_3} |\tilde{d}_{,tx_3}|_{3,S_2}) \|w\|_{1,\Omega} \|w_t\|_{1,\Omega}.$$

Next we consider J_1. Repeating the estimate of I_3 in (3.15), we have

$$|J_1| \leq c(\varrho^{1/6} + \varepsilon) \|\tilde{d}_t\|_{1,3,\Omega} \|w\|_{1,\Omega} \|w_t\|_{1,\Omega}.$$

The fourth integral in the first term on the r.h.s. of (3.47) equals

$$\int_\Omega \delta \cdot \nabla w_t \cdot w_t dx = \int_\Omega b \cdot \nabla w_t \cdot w_t dx + \int_\Omega \nabla \varphi \cdot \nabla w_t w_t dx \equiv J_3 + J_4.$$

Repeating estimate of I_1 from (3.11) we have

$$|J_3| \leq c\varrho^{1/6} \|\tilde{d}\|_{1,\Omega} \|w_t\|_{1,\Omega}^2,$$

and using estimate of I_2 yields

$$|J_4| \leq c(\varepsilon \varrho^{\mu-2/3} \sup_{x_3} |\tilde{d}|_{3,S_2} + \varrho^{\mu+1/3} \sup_{x_3} |\tilde{d}_{,x_3}|_{3,S_2}) \|w_t\|_{1,\Omega}^2.$$

Continuing, the fifth integral in the first term on the r.h.s. of (3.47) has the form

$$\int_\Omega \delta_t \cdot \nabla w \cdot w_t dx = \int_\Omega b_t \cdot \nabla w w_t dx + \int_\Omega \nabla \varphi_t \cdot \nabla w \cdot \nabla w_t dx \equiv J_5 + J_6.$$

Repeating the estimates of terms I_1 and I_2 from (3.11) we have

$$|J_5| \leq c\varrho^{1/6} \|\tilde{d}_t\|_{1,\Omega} \|w\|_{1,\Omega} \|w_t\|_{1,\Omega}$$

and

$$|J_6| \leq c[\varepsilon \varrho^{\mu-2/3} \sup_{x_3} |\tilde{d}_t|_{3,S_2} + \varrho^{\mu+1/3} \sup_{x_3} |\tilde{d}_{tx_3}|_{3,S_2}] \|w\|_{1,\Omega} \|w_t\|_{1,\Omega}.$$

Finally, the last integral from the first term on the r.h.s. of (3.47) takes the form

$$\int_\Omega w_t \cdot \nabla \delta \cdot w_t dx = \int_\Omega w_t \cdot \nabla b \cdot w_t dx + \int_\Omega w_t \cdot \nabla \nabla \varphi \cdot w_t dx \equiv J_7 + J_8.$$

Repeating the estimates of I_3 and I_4 from (3.14) we derive

$$|J_7| \le c(\varrho^{1/6} + \varepsilon)\|\tilde{d}\|_{1,3,\Omega}\|w_t\|_{1,\Omega}^2$$

and

$$|J_8| \le c[\varepsilon \varrho^{\mu-2/3} \sup_{x_3} |\tilde{d}|_{3,S_2} + \varrho^{\mu+1/3} \sup_{x_3} |\tilde{d}_{,x_3}|_{3,S_2}]\|w_t\|_{1,\Omega}^2.$$

Summarizing the above estimates gives

$$\left| \int_\Omega (w \cdot \nabla w_t + w_t \cdot \nabla w + w \cdot \nabla \delta_t + \delta \cdot \nabla w_t + \delta_t \cdot \nabla w + w_t \cdot \nabla \delta) \cdot w_t dx \right|$$

$$\le c(\varrho^{1/6} + \varepsilon)\|\tilde{d}_t\|_{1,3,\Omega}\|w\|_{1,\Omega}\|w_t\|_{1,\Omega} + c(\varrho^{1/6} + \varepsilon)\|\tilde{d}\|_{1,3,\Omega}\|w_t\|_{1,\Omega}^2$$

$$+ c\varrho^{1/6}\|\tilde{d}\|_{1,\Omega}\|w_t\|_{1,\Omega}^2 + c\varrho^{1/6}\|\tilde{d}_t\|_{1,\Omega}\|w\|_{1,\Omega}\|w_t\|_{1,\Omega} \tag{3.48}$$

$$+ c(\varepsilon \varrho^{\mu-2/3} \sup_{x_3} |\tilde{d}_{,t}|_{3,S_2} + \varrho^{\mu+1/3} \sup_{x_3} |\tilde{d}_{,tx_3}|_{3,S_2})\|w\|_{1,\Omega}\|w_t\|_{1,\Omega}$$

$$+ c(\varepsilon \varrho^{\mu-2/3} \sup_{x_3} |\tilde{d}|_{3,S_2} + \varrho^{\mu+1/3} \sup_{x_3} |\tilde{d}_{,x_3}|_{3,S_2})\|w_t\|_{1,\Omega}^2.$$

Next, we examine the third term on the r.h.s. of (3.47). Repeating the considerations leading to (3.17), we infer

$$\sum_{\alpha=1}^{2} |\delta_t \cdot \bar{\tau}_\alpha|_{2,S_1}^2 \le c\|\tilde{d}_t\|_{1,3/2,\Omega}^2 + c\frac{\varepsilon^2}{\varrho^{2/3}} \exp\left(\frac{2}{3\varepsilon}\right) \sup_{x_3} |\tilde{d}_t|_{3/2,S_2}^2. \tag{3.49}$$

In virtue of estimate (3.18) we have the following estimate for the second term on the r.h.s. of (3.47)

$$|\mathbb{D}(\delta_t)|_{2,\Omega}^2 \le c\sum_{i=1}^{2}[\|\tilde{d}_{i,t}\|_{1,2,\Omega}^2 + \frac{\varepsilon^2}{\varrho}e^{1/\varepsilon} \sup_{x_3} |\tilde{d}_{i,t}|_{2,S_2}^2]. \tag{3.50}$$

Analyzing the last integral on the r.h.s. of (3.47) one gets

$$\left| \int_\Omega (f_t - \delta_{tt} - \delta \cdot \nabla \delta_t - \delta_t \cdot \nabla \delta) \cdot w_t dx \right|$$

$$\le \varepsilon_2 |w_t|_{6,\Omega}^2 + c/\varepsilon_2(|f_t|_{6/5,\Omega}^2 + |\delta_{tt}|_{6/5,\Omega}^2) \tag{3.51}$$

$$+ \left| \int_\Omega (\delta \cdot \nabla \delta_t + \delta_t \cdot \nabla \delta) \cdot w_t dx \right|.$$

Repeating estimate (3.19) in this case we have

$$|\delta_{tt}|_{6/5,\Omega} \le \|\tilde{d}_{tt}\|_{1,6/5,\Omega} + c\frac{\varepsilon}{\varrho^{1/6}}e^{1/6\varepsilon}\sup_{x_3}|\tilde{d}_{tt}|_{6/5,S_2}. \tag{3.52}$$

Finally, in view of (3.20), we obtain

$$\left|\int_\Omega (\delta\cdot\nabla\delta_t + \delta_t\cdot\nabla\delta)\cdot w_t dx\right|$$
$$\le |\nabla\delta_t|_{2,\Omega}|\delta|_{3,\Omega}|w_t|_{6,\Omega} + |\nabla\delta|_{2,\Omega}|\delta_t|_{3,\Omega}|w_t|_{6,\Omega}$$
$$\le \varepsilon_3|w_t|_{6,\Omega}^2 + c(1/\varepsilon_3)\|\delta_t\|_{1,2,\Omega}^2\|\delta\|_{1,2,\Omega}^2$$
$$\le \varepsilon_3|w_t|_{6,\Omega}^2 + c(1/\varepsilon_3)\left(\|\tilde{d}\|_{1,2,\Omega}^2 + \frac{\varepsilon^2}{\varrho}e^{1/\varepsilon}\sup_{x_3}|\tilde{d}|_{2,S_2}^2\right)$$
$$\cdot\left(\|\tilde{d}_t\|_{1,2,\Omega}^2 + \frac{\varepsilon^2}{\varrho}e^{1/\varepsilon}\sup_{x_3}|\tilde{d}_t|_{2,S_2}^2\right). \tag{3.53}$$

Employing the above estimates in (3.47) yields

$$\frac{1}{2}\frac{d}{dt}|w_t|_{2,\Omega}^2 + \nu\|w_t\|_{1,\Omega}^2 + \gamma\sum_{\alpha=1}^{2}|w_t\cdot\bar{\tau}_\alpha|_{2,S_1}^2 \le c|w|_{\infty,\Omega}^2|w_t|_{2,\Omega}^2$$
$$+ c[(\varrho^{1/6} + \varepsilon)\|\tilde{d}_t\|_{1,3,\Omega}\|w\|_{1,\Omega}\|w_t\|_{1,\Omega}$$
$$+ (\varrho^{1/6} + \varepsilon)\|\tilde{d}\|_{1,3,\Omega}\|w_t\|_{1,\Omega}^2$$
$$+ \varrho^{1/6}\|\tilde{d}\|_{1,\Omega}\|w_t\|_{1,\Omega}^2 + c\varrho^{1/6}\|\tilde{d}_t\|_{1,\Omega}\|w\|_{1,\Omega}\|w_t\|_{1,\Omega}$$
$$+ (\varepsilon\varrho^{\mu-2/3}\sup_{x_3}|\tilde{d}_t|_{3,S_2} + \varrho^{\mu+1/3}\sup_{x_3}|\tilde{d}_{,tx_3}|_{3,S_2})\|w\|_{1,\Omega}\|w_t\|_{1,\Omega}$$
$$+ (\varepsilon\varrho^{\mu-2/3}\sup_{x_3}|\tilde{d}|_{3,S_2} + \varrho^{\mu+1/3}\sup_{x_3}|\tilde{d}_{,x_3}|_{3,S_2})\|w_t\|_{1,\Omega}^2] \tag{3.54}$$
$$+ c[|f_t|_{6/5,\Omega}^2 + \|\tilde{d}_{tt}\|_{1,6/5,\Omega}^2 + \frac{\varepsilon}{\varrho^{1/6}}e^{1/6\varepsilon}\sup_{x_3}|\tilde{d}_{tt}|_{6/5,S_2}^2$$
$$+ \|\tilde{d}_t\|_{1,3/2,\Omega}^2 + \frac{\varepsilon^2}{\varrho^{2/3}}e^{2/3\varepsilon}\sup_{x_3}|\tilde{d}_t|_{3/2,S_2}^2 + \|\tilde{d}_{,t}\|_{1,2,\Omega}^2$$
$$+ \frac{\varepsilon^2}{\varrho}e^{1/\varepsilon}\sup_{x_3}|\tilde{d}_{,t}|_{2,S_2}^2 + \left(\|\tilde{d}\|_{1,2,\Omega}^2 + \frac{\varepsilon^2}{\varrho}e^{1/\varepsilon}\sup_{x_3}|\tilde{d}|_{2,S_2}^2\right)$$
$$\cdot\left(\|\tilde{d}_t\|_{1,2,\Omega}^2 + \frac{\varepsilon^2}{\varrho}e^{1/\varepsilon}\sup_{x_3}|\tilde{d}_t|_{2,S_2}^2\right)].$$

We estimate the first square bracket on the r.h.s. of (3.54) by

$$c(\varepsilon\varrho^{\mu-2/3} + \varrho^{\mu+1/3} + \varrho^{1/6} + \varepsilon)[\|\tilde{d}\|_{W^1_{3,\infty}(\Omega)}\|w_t\|^2_{1,\Omega}$$

$$+\|\tilde{d}_t\|_{W^1_{3,\infty}(w)}\|w\|_{1,\Omega}\|w_t\|_{1,\Omega}].$$

The terms under the second square bracket on the r.h.s. of (3.54) are estimated by

$$c[|f_t|^2_{6/5,\Omega} + \|\tilde{d}_{tt}\|^2_{1,6/5,\Omega} + \|\tilde{d}_t\|^2_{1,2,\Omega} + \|\tilde{d}\|^4_{1,2,\Omega} + \|\tilde{d}_t\|^4_{1,2,\Omega}]$$

$$+c\left[\frac{\varepsilon}{\varrho^{1/6}}e^{1/6\varepsilon}\sup_{x_3}|\tilde{d}_{tt}|^2_{6/5,S_2} + \frac{\varepsilon^2}{\varrho^{2/3}}e^{2/3\varepsilon}\sup_{x_3}|\tilde{d}_t|^2_{2,S_2} + \frac{\varepsilon^2}{\varrho}e^{1/\varepsilon}\sup_{x_3}|\tilde{d}_t|^2_{2,S_2}\right.$$

$$\left.+\frac{\varepsilon^4}{\varrho^2}e^{2/\varepsilon}\sup_{x_3}|\tilde{d}|^4_{2,S_2} + \frac{\varepsilon^4}{\varrho^2}e^{2/\varepsilon}\sup_{x_3}|\tilde{d}_t|^4_{2,S_2}\right].$$

Using that ε small and $\varrho < 1$ inequality (3.54) simplifies to

$$\frac{1}{2}\frac{d}{dt}|w_t|^2_{2,\Omega} + \nu\|w_t\|^2_{1,\Omega} + \gamma\sum_{\alpha=1}^2 |w_t \cdot \bar{\tau}_\alpha|^2_{2,S_1} \le c|w|^2_{\infty,\Omega}|w_t|^2_{2,\Omega}$$

$$+ c(\varepsilon\varrho^{\mu-2/3} + \varrho^{\mu+1/3} + \varrho^{1/6} + \varepsilon)[\|\tilde{d}\|_{W^1_{3,\infty}(\Omega)}\|w_t\|^2_{1,\Omega}$$

$$+ \|\tilde{d}_t\|_{W^1_{3,\infty}(\Omega)}\|w\|_{1,\Omega}\|w_t\|_{1,\Omega}]$$

$$+ c[|f_t|^2_{6/5,\Omega} + \|\tilde{d}_{tt}\|^2_{1,6/5,\Omega} + \|\tilde{d}_t\|^2_{1,2,\Omega} + \|\tilde{d}\|^4_{1,2,\Omega} + \|\tilde{d}_t\|^4_{1,2,\Omega}] \quad (3.55)$$

$$+ c\left[\frac{\varepsilon}{\varrho^{1/6}}e^{1/6\varepsilon}\sup_{x_3}|\tilde{d}_{tt}|^2_{6/5,S_2} + \frac{\varepsilon^2}{\varrho}e^{1/\varepsilon}\sup_{x_3}|\tilde{d}_t|^2_{2,S_2}\right.$$

$$\left.+ \frac{\varepsilon^4}{\varrho^2}e^{2/\varepsilon}\sup_{x_3}|\tilde{d}|^4_{2,S_2} + \frac{\varepsilon^4}{\varrho^2}e^{2/\varepsilon}\sup_{x_3}|\tilde{d}_t|^4_{2,S_2}\right].$$

Now, we simplify (3.55). Assume that $\mu > \frac{2}{3}$ and ϱ will be chosen small. Then the second term on the r.h.s. of (3.55) is bounded by

$$c(\varrho^{1/6} + \varepsilon)(\|\tilde{d}\|_{W^1_{3,\infty}(\Omega)} + \|\tilde{d}_t\|_{W^1_{3,\infty}(\Omega)})(\|w_t\|^2_{1,\Omega} + \|w\|^2_{1,\Omega}).$$

Next, we assume that

$$c(\varrho^{1/6} + \varepsilon)(\|\tilde{d}\|_{W^1_{3,\infty}(w)} + \|\tilde{d}_t\|_{W^1_{3,\infty}(\Omega)}) \le \frac{\nu}{k_0}, \quad (3.56)$$

where k_0 will be chosen sufficiently large. Introducing the notation

$$\alpha = \|\tilde{d}\|_{W^1_{3,\infty}(\Omega)} + \|\tilde{d}_t\|_{W^1_{3,\infty}(\Omega)},$$

we obtain from (3.56) that

$$\varrho^{1/6} \le \frac{c}{\alpha}, \quad \varepsilon \le \frac{c}{\alpha}. \tag{3.57}$$

To examine the last term on the r.h.s. of (3.55), we calculate

$$\frac{\varepsilon}{\varrho^{1/6}} e^{1/6\varepsilon} \le ce^{c\alpha},$$

$$\frac{\varepsilon^2}{\varrho} e^{1/\varepsilon} \le \frac{(c/\alpha)^2}{(c/\alpha)^6} e^{c\alpha} = c\alpha^4 e^{c\alpha}$$

$$\frac{\varepsilon^4}{\varrho^2} e^{2/\varepsilon} \le \frac{(c/\alpha)^4}{(c/\alpha)^{12}} e^{c\alpha} = c\alpha^8 e^{c\alpha}.$$

In view of the above considerations we obtain from (3.55) the inequality

$$\frac{d}{dt}|w_t|^2_{2,\Omega} + \nu\|w_t\|^2_{1,\Omega} + \gamma\sum_{\alpha=1}^{2}|w_t \cdot \bar{\tau}_\alpha|^2_{2,S_1}$$

$$\le c|w|^2_{\infty,\Omega}|w_t|^2_{2,\Omega} + \frac{\nu}{k}\|w\|^2_{1,\Omega}$$

$$+ c[|f|^2_{6/5,\Omega} + |f_t|^2_{6/5,\Omega} + \|\tilde{d}_{tt}\|^2_{1,6/5,\Omega} + \|\tilde{d}_t\|^2_{1,2,\Omega} + \|\tilde{d}\|^4_{1,2,\Omega} \tag{3.58}$$

$$+ \|\tilde{d}_t\|^4_{1,2,\Omega}] + c\varphi(\|\tilde{d}\|_{W^1_{3,\infty}(\Omega)}, \|\tilde{d}_t\|_{W^1_{3,\infty}(\Omega)}) \cdot [|\tilde{d}_{tt}|^2_{6/5,\infty,\Omega}$$

$$+ |\tilde{d}_t|^2_{2,\infty,\Omega} + |\tilde{d}|^4_{2,\infty,\Omega} + |\tilde{d}_t|^4_{2,\infty,\Omega} + \|\tilde{d}\|_{W^1_{3,\infty}(\Omega)} + \|\tilde{d}_t\|^2_{1,6/5,\Omega}]$$

$$\equiv c|w|^2_{\infty,\Omega}|w_t|^2_{2,\Omega} + \frac{\nu}{k}\|w\|^2_{1,\Omega} + \Gamma^2(t).$$

From (3.27) and (3.58) we have

$$\frac{d}{dt}(|w|^2_{2,\Omega} + |w_t|^2_{2,\Omega}) + \nu(\|w\|^2_{1,\Omega} + \|w_t\|^2_{1,\Omega})$$

$$\le c|w|^2_{\infty,\Omega}|w_t|^2_{2,\Omega} + F^2(t) + \Gamma^2(t). \tag{3.59}$$

From (3.59) it follows

$$\frac{d}{dt}(|w|^2_{2,\Omega} + |w_t|^2_{2,\Omega}) + \nu(|w|^2_{2,\Omega} + |w_t|^2_{2,\Omega})$$

$$\le c|w|^2_{\infty,\Omega}(|w|^2_{2,\Omega} + |w_t|^2_{2,\Omega}) + F^2(t) + \Gamma^2(t). \tag{3.60}$$

Hence, one obtains

$$\frac{d}{dt}\left[(|w|_{2,\Omega}^2 + |w_t|_{2,\Omega}^2)\exp\left(\nu t - \int_{kT}^t |w(t')|_{\infty,\Omega}^2 dt'\right)\right]$$

$$\leq [F^2(t) + \Gamma^2(t)]\exp\left(\nu t - \int_{kT}^t |w(t')|_{\infty,\Omega}^2 dt'\right). \tag{3.61}$$

Integrating with respect to time from kT to $t \in (kT, (k+1)T]$, $k \in \mathbb{N}_0$, we derive

$$|w(t)|_{2,\Omega}^2 + |w_t(t)|_{2,\Omega}^2 \leq \exp\left(\int_{kT}^t |w(t')|_{\infty,\Omega}^2 dt'\right)$$

$$\cdot \int_{kT}^t (F^2(t') + \Gamma^2(t'))dt' + \exp\left[-\nu(t-kT) + \int_{kT}^t |w(t')|_\infty^2 dt'\right] \tag{3.62}$$

$$\cdot(|w(kT)|_{2,\Omega}^2 + |w_t(kT)|_{2,\Omega}^2).$$

Setting $t = (k+1)T$, introducing the notation

$$B_1^2(T) = \sup_{k\in\mathbb{N}_0} \exp\left(\int_{kT}^{(k+1)T} |w(t)|_{\infty,\Omega}^2 dt\right) \cdot \int_{kT}^{(k+1)T} (F^2(t) + \Gamma^2(t))dt, \tag{3.63}$$

and assuming that the following restriction holds, with A_2 defined in (3.34)

$$\int_{kT}^{(k+1)T} |w(t)|_{\infty,\Omega}^2 dt \leq cT^{1/4} A_2^{1/2}(T)\left(\int_{kT}^{(k+1)T} |w_{xx}(t)|_{2,\Omega}^2 dt\right)^{3/4} \tag{3.64}$$

$$\leq \frac{\nu}{2}T \quad \text{for any } k \in \mathbb{N}_0,$$

we obtain from (3.62) the inequality

$$|w((k+1)T)|_{2,\Omega}^2 + |w_t((k+1)T)|_{2,\Omega}^2 \leq B_1^2(T)$$

$$+ \exp\left(-\frac{\nu}{2}T\right)(|w(kT)|_{2,\Omega}^2 + |w_t(kT)|_{2,\Omega}^2). \tag{3.65}$$

By iteration, we have

$$|w(kT)|_{2,\Omega}^2 + |w_t(kT)|_{2,\Omega}^2 \leq \frac{B_1^2(T)}{1 - \exp\left(-\frac{\nu}{2}T\right)}$$

$$+ \exp(-\nu kT)(|w(0)|_{2,\Omega}^2 + |w_t(0)|_{2,\Omega}^2). \tag{3.66}$$

In view of (3.63) and (3.64) we obtain

$$|w(t)|^2_{2,\Omega} + |w_t(t)|^2_{2,\Omega}$$

$$\leq B_1^2(T) + \exp\left(-\frac{\nu}{2}T\right)(|w(kT)|^2_{2,\Omega} + |w_t(kT)|^2_{2,\Omega}). \tag{3.67}$$

Using (3.66) in (3.67) we obtain for $t \in (kT, (k+1)T]$ the inequality

$$|w(t)|^2_{2,\Omega} + |w_t(t)|^2_{2,\Omega} \leq \frac{B_1^2(T)}{1 - \exp\left(-\frac{\nu}{2}T\right)} \tag{3.68}$$

$$+ \exp(-\nu kT)(|w(0)|^2_{2,\Omega} + |w_t(0)|^2_{2,\Omega}).$$

Integrating (3.59) with respect to time from $t = kT$ to $t \in (kT, (k+1)T]$ and using (3.68), we obtain

$$\|w\|^2_{V(\Omega \times (kT,t))} + \|w_t\|^2_{V(\Omega \times (kT,t))}$$

$$\leq \left[1 + \int_{kT}^{(k+1)T} |w(t)|^2_{\infty,\Omega} dt\right]\left[\frac{B_1^2}{1 - \exp\left(-\frac{\nu}{2}T\right)}\right. \tag{3.69}$$

$$\left. + \exp(-\nu kT)(|w(0)|^2_{2,\Omega} + |w_t(0)|^2_{2,\Omega})\right] + c \int_{kT}^{(k+1)T} (F^2(t) + \Gamma^2(t)dt).$$

The above considerations imply the following result.

Lemma 3.3 *Assume that $F, \Gamma \in L_2(kT, (k+1)T)$ for any $k \in \mathbb{N}_0$, where F is defined in (3.27) and Γ in (3.58). Assume that $w \in L_2(kT, (k+1)T; L_\infty(\Omega))$ and restriction (3.64) holds, with initial data such that $w(0), w_t(0) \in L_2(\Omega)$. Then estimates (3.68) and (3.69) hold for any $k \in \mathbb{N}_0$, where $B_1(T)$ is defined in (3.63).*

Having estimate (3.69) for w and w_t we can derive an estimate for v and v_t.

Proposition 3.4 *Let the assumptions of Lemmas 3.2 and 3.3 hold. Then*

$$\|v\|_{V(\Omega \times (kT,t))} \leq c\|w\|_{V(\Omega \times (kT,t))} + F_1^2(kT, t), \tag{3.70}$$

where $F_1(kT, t)$ is defined in the property (2) of Lemma 3.2 as a quantity $\varphi(\sup_t \|d(t)\|_{1,3,S_2}) \cdot \|d\|^2_{\frac{1}{2},2,S_2 \times (kT,t)}$. Next

$$\|v_t\|_{V(kT,t)} \leq \|w_t\|_{V(kT,t)} + \|\delta_t\|_{V(kT,t)}, \tag{3.71}$$

where

$$\|\delta_t\|_{V(kT,t)} \leq \varphi(\sup_t \|d(t)\|_{1,3,S_2}, \sup_t \|d_t(t)\|_{1,3,S_2})$$

$$\cdot(\|d\|_{1/2,2,S_2 \times (kT,t)} + \|d_t\|_{1/2,2,S_2 \times (kT,t)}) \equiv \Gamma_1^2(kT,t).$$

Proof From the definition of δ_t we have

$$\|\delta_t\|_{V(\Omega \times (kT,t))}^2 \leq \|b_t\|_{V(\Omega \times (kT,t))}^2 + \|\nabla\varphi_t\|_{V(\Omega \times (kT,t))}^2$$

$$\leq |b_t|_{2,\infty,\Omega \times (kT,t)}^2 + \|b_t\|_{1,2,2,\Omega \times (kT,t)}^2 + |\nabla\varphi_t|_{2,\infty,\Omega \times (kT,t)}^2$$

$$+\|\nabla\varphi_t\|_{1,2,2,\Omega \times (kT,t)}^2 \equiv J.$$

Now we estimate the particular terms in J. Repeating the considerations leading to estimate I and (3.57) we obtain that $J \leq \Gamma_1^2$. This concludes the proof.   □

Chapter 4
Local Estimates for Regular Solutions

Abstract The aim of this chapter is to show the estimate

$$\|v\|_{W_2^{2,1}(\Omega^t)} \leq M,$$

where M depends on some norms of data but does not depend on time $t \in (0, T]$ explicitly. The dependence on time is only through integrals with respect to time of data functions f, d and their time and space derivatives. To prove the above inequality, we need smallness of the following quantity

$$\Lambda_2(t) = \int_0^t \left(\|d_{x'}\|_{H^1(S_2)}^2 + \|d_t\|_{H^1(S_2)}^2 + |f_3|_{L_{4/3}(S_2)}^2 + |g|_{L_{6/5}(\Omega)}^2 \right) dt'$$

$$+ \mathcal{A} \sup_t \|d_{x'}\|_{W_{3/2}^1(S_2)}^2 + |h(0)|_{L_2(\Omega)}^2,$$

where $h = v_{,x_3}, g = f_{,x_3}$, and $\mathcal{A}$ estimates the weak solution (see Chap. 3). To achieve this, we consider the problems for variables $h, q = p_{,x_3}$ and $\chi = v_{2,x_1} - v_{1,x_2}$. Then for sufficiently small $\Lambda_2(t)$ we can apply some fixed point argument and use energy estimates for solutions to problems on h, q and χ to obtain the desired bound on v by M.

The main parts of this monograph are this chapter and Chap. 5. In this chapter the finite time existence of solutions to problem (1.1) is established. The existence time is inversely proportional to the quantity Λ_2. In this chapter we are not interested in the length of T but we perform all estimates in such a way that in fact there is no restriction on magnitude of T. Here we derive an a priori estimate and with this we are able to prove the existence of regular solutions in Chap. 12, Sect. 12.2 by the method of the Leray-Schauder theorem. On the other hand, to achieve the estimate in this chapter we need the estimate and existence of weak solutions (see Chap. 3).

To infer the estimate for weak solutions in Chap. 3, we had to apply some delicate considerations implied by the fact that the inflow-outflow problem

© Springer Nature Switzerland AG 2019

J. Renclawowicz, W. M. Zajączkowski, *The Large Flux Problem to the Navier-Stokes Equations*, Advances in Mathematical Fluid Mechanics, https://doi.org/10.1007/978-3-030-32330-1_4

was analyzed. Then, the corresponding result was shown in Lemma 3.2, in Chap. 3.

The first result in this chapter is the energy type inequality for $\|h\|_{V(\Omega^T)}$ (see (4.42)). To prove this inequality we need the existence and estimates for Green functions to problem (4.19) and (4.20) which are proved in Lemmas 2.11 and 2.12.

Since Ω is the cylindrical domain parallel to the x_3-axis we can introduce vorticity $\chi(x,t)$ which helps us to increase the regularity of the first two components of velocity, i.e., $v' = (v_1, v_2)$ (see Lemma 4.9 and inequality (4.56)). We next use the following estimate for the Stokes system (see [Z8], Theorem A), for $\sigma \in (1, \infty)$.

$$
\begin{aligned}
&\|v\|_{W^{2,1}_\sigma(\Omega^T)} + \|\nabla p\|_{L_\sigma(\Omega^T)} \\
&\leq c(|\bar{f}|_{\sigma, \Omega^T} + \|d\|_{W^{2-1/\sigma, 1-1/2\sigma}_\sigma(S^T_2)} + \|v(0)\|_{W^{2-2/\sigma}_\sigma(\Omega)}),
\end{aligned}
\tag{4.1}
$$

where the Stokes system has the form

$$
\begin{aligned}
v_{,t} - \operatorname{div} \mathbb{T}(v,p) = -v' \cdot \nabla v - v_3 h + f \equiv \bar{f} &\quad \text{in } \Omega^T, \\
\operatorname{div} v = 0 &\quad \text{in } \Omega^T, \\
v \cdot \bar{n} = 0, \quad \nu\bar{n} \cdot \mathbb{D}(v) \cdot \bar{\tau}_\alpha + \gamma v \cdot \bar{\tau}_\alpha = 0, \quad \alpha, 1, 2 &\quad \text{on } S^T_1, \\
v \cdot \bar{n} = d, \quad \bar{n} \cdot \mathbb{D}(v) \cdot \bar{\tau}_\alpha = 0, \quad \alpha = 1, 2 &\quad \text{on } S^T_2, \\
v|_{t=0} = v(0) &\quad \text{in } \Omega.
\end{aligned}
$$

Combining this with inequality (4.56) we conclude (see (4.62))

$$
\|v\|_{W^{2,1}_{5/3}(\Omega^T)} \leq \varphi(\mathcal{A})H_2 + \varphi(D),
\tag{4.2}
$$

where D describes all data norms, $\mathcal{A}$ is the estimate for the weak solution given in (4.13), and

$$
H_2 = H_1 + |h|_{10/3, \Omega^t}
$$
$$
H_1 = \sup_t |h|_{3, \Omega} + \|h\|_{1, 2, \Omega^t}.
$$

For solutions to problem (4.6) we derive the estimate

$$
\|h\|_{V(\Omega^t)} \leq \varphi(D_1, \mathcal{V})\Lambda_2(t),
$$

where $D_1(t) = |d_1|_{3, 6, S^t}, \mathcal{V}(t) = |\nabla v|_{3, 2, \Omega^t}$. Then for the problem (4.9) for two-dimensional vorticity we obtain inequality (4.54)

$$\|\chi\|_{V(\Omega^t)}^2 \le c(1 + \mathcal{A}^2)(|h_3|_{3,\infty,\Omega^t}^2 + |F_3|_{6/5,2,\Omega^t}^2$$
$$+\|v'\|_{W_2^{1,1/2}(\Omega^t)}^2 + \|v'\|_{5/6,2,\infty,\Omega^t}^2 + |\chi(0)|_{2,\Omega}^2).$$

Next the rot-div problem (4.57) implies the inequality (4.60)

$$\|v'\|_{V^1(\Omega^t)}^2 \le \varphi(D_1, \mathcal{V})\Lambda_2^2(t)$$
$$+c(1 + \mathcal{A}^2)(|h_3|_{3,\infty,\Omega^t}^2 + \|h_3\|_{V(\Omega^t)}^2 + |F_3|_{6/5,2,\Omega^t}^2$$
$$+|\chi(0)|_{2,\Omega}^2 + \|v'\|_{W_2^{1,1/2}(\Omega^t)}^2 + \|v'\|_{5/6,2,\infty,\Omega^t}^2).$$

Consider the Stokes problem (4.63) where the nonlinear term $v \cdot \nabla v$ is expressed in the form

$$v' \cdot \nabla' v + v_3 \cdot h$$

and estimated by (4.64), (4.65) as follows

$$|v \cdot \nabla v|_{5/3,\Omega^T} \le |v' \cdot \nabla v|_{5/3,\Omega^T} + |v_3 h|_{5/3,\Omega^T} \le \mathcal{A}(\|v'\|_{V^1(\Omega^T)} + |h|_{10/3,\Omega^T}).$$

Then, for sufficiently small Λ_2 we obtain the estimate (see (4.62) and Lemma 4.10)

$$\|v\|_{W_{5/3}^{2,1}(\Omega^T)} \le M,$$

where M depends on H_2 and data. Using this for the Stokes problem (4.63) we increase regularity of v to obtain (see (4.73))

$$\|v\|_{W_2^{2,1}(\Omega^T)} \le \Phi(H_2, D), \tag{4.3}$$

where D contains norms of data functions. Using for $\frac{5}{3} \le \sigma \le \frac{10}{3}$ the imbedding

$$H_2 \le c\|h\|_{W_\sigma^{2,1}(\Omega^T)}$$

and applying (4.3) and (4.1) for the problem (4.6) we attain, for sufficiently small parameter Λ_2, the estimate (4.75) of the form:

$$\|h\|_{W_\sigma^{2,1}(\Omega^T)} \le cD_9(t), \quad \frac{5}{3} \le \sigma \le \frac{10}{3}, \tag{4.4}$$

where D_9 is defined by (4.77). This yields (4.4) and the estimate for $v \in W_2^{2,1}(\Omega^T)$ (see Theorem 4.14) for sufficiently small $\Lambda_2(t)$ and finite time

t. Let us note that although T in this theorem can be chosen arbitrarily large, on the other hand we require smallness of $\Lambda_2(T)$, where

$$\Lambda_2^2(T) = \int_0^T (\|d_{x'}\|_{1,S_2}^2 + \|d_t\|_{1,S_2}^2 + |f_3|_{4/3,S_2}^2 + |g|_{6/5,\Omega}^2)dt$$

$$+ \mathcal{A} \sup_t \|d_{x'}\|_{1,3/2,S_2}^2 + |h(0)|_{2,\Omega}^2.$$

If we look closely at this formula, we observe that for large T the necessity of smallness implies stronger decay restrictions in norms integrated in time T. Therefore, we need some balance between small $\Lambda_2(T)$ and possibly large, but not too large T, in order to satisfy the assumptions for global existence. This is the topic of Chap. 5.

Proceeding into details, to obtain a higher regularity for solutions to problem (1.1) we introduce the quantities

$$h = v_{,x_3}, \quad q = p_{,x_3}, \quad g = f_{,x_3}. \tag{4.5}$$

Lemma 4.1 *Functions h, q defined in (4.5) are solutions to the system:*

$$\begin{aligned}
h_t - \operatorname{div} \mathbb{T}(h,q) &= -v \cdot \nabla h - h \cdot \nabla v + g && \text{in } \Omega^T, \\
\operatorname{div} h &= 0 && \text{in } \Omega^T, \\
h \cdot \bar{n} &= 0 && \text{on } S_1^T, \\
\nu \bar{n} \cdot \mathbb{D}(h) \cdot \bar{\tau}_\alpha + \gamma h \cdot \bar{\tau}_\alpha &= 0, \quad \alpha = 1, 2, && \text{on } S_1^T, \\
h_i &= -d_{,x_i}, \quad i = 1, 2 && \text{on } S_2^T, \\
h_{3,x_3} &= \Delta' d && \text{on } S_2^T, \\
h|_{t=0} &= h(0) && \text{in } \Omega,
\end{aligned} \tag{4.6}$$

where $\Delta' = \partial_{x_1}^2 + \partial_{x_2}^2$.

Proof Equations $(4.6)_{1,2,3,4,7}$ follow directly from the corresponding equations in (1.1) by differentiation with respect to x_3, because S_1 is parallel to the x_3-axis. To show $(4.6)_{5,6}$ we recall that

$$v_3|_{S_2} = d, \quad (v_{i,x_3} + v_{3,x_i})|_{S_2} = 0, \quad i = 1, 2. \tag{4.7}$$

Hence, $v_{i,x_3} = -d_{,x_i}$, $i = 1, 2$ and $(4.6)_5$ holds. From $(1.1)_2$ we have that $v_{3,x_3x_3}|_{S_2} = -(v_{1,x_1x_3} + v_{2,x_2x_3})|_{S_2} = d_{,x_1x_1} + d_{,x_2x_2} = \Delta' d$ and $(4.6)_6$ follows. This ends the proof. $\square$

Lemma 4.2 *The function*

$$\chi = v_{2,x_1} - v_{1,x_2} \tag{4.8}$$

is a solution to the problem

$$\chi_t + v \cdot \nabla\chi - h_3\chi + h_2 w_{,x_1} - h_1 w_{,x_2} - \nu\Delta\chi = F_3 \quad\text{in } \Omega^T,$$

$$\chi = -v_i(n_{i,x_j}\tau_{1j} + \tau_{1i,x_j}n_j) + \gamma v v_j\tau_{1j}$$

$$\quad + v \cdot \bar{\tau}_1(\tau_{12,x_1} - \tau_{11,x_2}) \equiv \chi_* \qquad\text{on } S_1^T, \tag{4.9}$$

$$\chi_{,x_3} = 0 \qquad\text{on } S_2^T,$$

$$\chi|_{t=0} = \chi(0) \qquad\text{in } \Omega,$$

where $F_3 = f_{2,x_1} - f_{1,x_2}$, $w = v_3$ *and the tangent and normal vectors have the form*

$$\bar{n}|_{S_1} = \frac{(\varphi_{0,x_1}, \varphi_{0,x_2}, 0)}{\sqrt{\varphi_{0,x_1}^2 + \varphi_{0,x_2}^2}}, \quad \bar{\tau}_1|_{S_1} = \frac{(-\varphi_{0,x_2}, \varphi_{0,x_1}, 0)}{\sqrt{\varphi_{0,x_1}^2 + \varphi_{0,x_2}^2}}, \quad \bar{\tau}_2|_{S_1} = (0,0,1) \equiv \bar{e}_3,$$

$$\bar{n}|_{S_2(a_i)} = (-1)^i \bar{e}_3, \quad i = 1,2, \quad a_1 = -a, a_2 = a, \tag{4.10}$$

$$\bar{\tau}_1|_{S_2} = (1,0,0) \equiv \bar{e}_1, \quad \bar{\tau}_2|_{S_2} = (0,1,0) \equiv \bar{e}_2.$$

Proof Differentiating the first equation of $(1.1)_1$ with respect to x_2, the second equation of $(1.1)_1$ with respect to x_1 and subtracting yields $(4.9)_1$. To show $(4.9)_2$ we extend vectors $\bar{\tau}_1$, $\bar{n}$ into a neighborhood of S_1. In this neighborhood $v' = (v_1, v_2)$ can be expressed in the form

$$v' = v \cdot \bar{\tau}_1\bar{\tau}_1 + v \cdot \bar{n}\bar{n}.$$

Then

$$\chi|_{S_1} = [(v \cdot \bar{\tau}_1\tau_{12} + v \cdot nn_2)_{,x_1} - (v \cdot \bar{\tau}_1\tau_{11} + v \cdot \bar{n}n_1)_{,x_2}]|_{S_1}$$

$$= [-\bar{n} \cdot \nabla(v \cdot \bar{\tau}_1) + v \cdot \bar{\tau}_1(\tau_{12,x_1} - \tau_{11,x_2})]|_{S_1}, \tag{4.11}$$

where we used that $v \cdot \bar{n}|_{S_1} = 0$ *and*

$$n_2\partial_{x_1}(v \cdot \bar{n}) - n_1\partial_{x_2}(v \cdot \bar{n}) = (n_2\partial_{x_1} - n_1\partial_{x_2})(v \cdot \bar{n}) = -(\tau_{11}\partial_{x_1} + \tau_{12}\partial_{x_2})(\bar{v} \cdot \bar{n}) = 0.$$

Utilizing $(1.1)_3$ in $(1.1)_4$ for $\alpha = 1$ yields

$$\nu\bar{n} \cdot \nabla(v \cdot \bar{\tau}_1) - \nu v_i(n_{i,x_j}\tau_{1j} + \tau_{1i,x_j}n_j) + \gamma v \cdot \bar{\tau}_1 = 0. \tag{4.12}$$

Exploiting (4.12) in (4.11) yields (4.9)$_2$. By the definition of χ and (4.6)$_5$ we have

$$\chi_{,x_3}|_{S_2} = (v_{2,x_1x_3} - v_{1,x_2x_3})|_{S_2} = -(d_{,x_1x_2} - d_{,x_2x_1})|_{S_2} = 0.$$

This ends the proof.   □

From Chap. 3 we have

$$\|v\|^2_{V(\Omega^t)} \le \varphi(\sup_t \|d\|_{1,3,S_2}) \int_0^t (|f|^2_{6/5,\Omega} + \|d\|^2_{1,3,S_2}$$

$$+ \|d_t\|^2_{1,6/5,S_2})dt' + |v(0)|^2_{2,\Omega} \equiv \mathcal{A}^2(t),$$
(4.13)

where φ is an increasing positive function.

4.1 A Priori Estimates for Function $h = v_{,x_3}$

To obtain the energy estimate for solutions to problem (4.6), we have to make the Dirichlet boundary condition on S_2^T homogeneous. For this purpose, we are looking for a function $\tilde{h}$ such that

$$\begin{aligned}
\operatorname{div} \tilde{h} &= 0 & &\text{in } \Omega, \\
\tilde{h} &= 0 & &\text{on } S_1, \\
\tilde{h}_i &= -d_{,x_i}, \quad i = 1, 2, & &\text{on } S_2, \\
\tilde{h}_3 &= 0 & &\text{on } S_2.
\end{aligned}$$
(4.14)

Lemma 4.3 *Assume that* $d = (d_1, d_2)$, $d_{,x'} \in W^1_\sigma(S_2)$, $d_{,x't} \in L_\sigma(S_2)$, $\sigma \in (1, \infty)$. *Then there exists a solution to problem (4.14) such that* $\tilde{h} \in W^1_\sigma(\Omega)$, $\tilde{h}_{,t} \in L_\sigma(\Omega)$, *and*

$$\|\tilde{h}\|_{1,\sigma,\Omega} \le c\|d_{,x'}\|_{1,\sigma,S_2},$$

$$|\tilde{h}_{,t}|_{\sigma,\Omega} \le c|d_{,x't}|_{\sigma,S_2}.$$
(4.15)

Proof First we define the functions

$$\bar{h}_i = -(\bar{\eta}_1 d_{1,x_i} + \bar{\eta}_2 d_{2,x_i}), \quad i = 1, 2,$$

$$\bar{h}_3 = 0,$$
(4.16)

where $\bar{\eta}_i = \bar{\eta}_i(x_3)$, $i = 1, 2$, are smooth cut-off functions such that $\bar{\eta}_1 = 1$, $\bar{\eta}_2 = 0$ near $S_2(-a)$ and $\bar{\eta}_1 = 0$, $\bar{\eta}_2 = 1$ near $S_2(a)$.

$S_2(a_i, \varrho)$, $i = 1, 2$, are introduced above (3.4). We have the compatibility condition

$$\sum_{i=1}^{2} n_i|_{S_1} d_{\alpha, x_i} = 0, \quad \alpha = 1, 2, \tag{4.17}$$

because $\bar{n}|_{S_1}$ does not depend on x_3, so

$$v \cdot \bar{n}|_{S_1} = 0 \Rightarrow v_{,x_3} \cdot \bar{n}|_{S_1} = 0 \Rightarrow h \cdot \bar{n}|_{S_1} = 0 \Rightarrow h|_{S_2} \cdot \bar{n}|_{S_1} = 0,$$

and (4.16) follows from the restriction $n_3|_{S_1} = 0$. Hence, $\bar{h}$ is a solution to the problem

$$
\begin{aligned}
&\operatorname{div} \bar{h} = -(\bar{\eta}_1 \Delta' d_1 + \bar{\eta}_2 \Delta' d_2) && \text{in } \Omega, \\
&\bar{h} \cdot \bar{\tau}_1 = -(\bar{\eta}_1 \tilde{d}_{1,x_j} + \bar{\eta}_2 \tilde{d}_{2,x_j}) \tau_{1j} && \text{on } S_1, \\
&\bar{h} \cdot \bar{\tau}_2 = 0 && \text{on } S_1, \\
&\bar{h} \cdot \bar{n} = 0 && \text{on } S_1, \\
&\bar{h}_i = -d_{j,x_i}, \quad i = 1, 2, && \text{on } S_2(a_j), \quad j = 1, 2, \\
&\bar{h}_3 = 0 && \text{on } S_2,
\end{aligned}
\tag{4.18}
$$

where the tangent and normal vectors are introduced in (4.10).

To construct function $\tilde{h}$ satisfying (4.14) we define function ϕ such that

$$
\begin{aligned}
&\Delta \phi = -(\bar{\eta}_1 \Delta' d_1 + \bar{\eta}_2 \Delta' d_2) && \text{in } \Omega, \\
&\bar{n} \cdot \nabla \phi = 0 && \text{on } S,
\end{aligned}
\tag{4.19}
$$

and functions λ and σ such that

$$
\begin{aligned}
&-\Delta \lambda + \nabla \sigma = 0 && \text{in } \Omega, \\
&\operatorname{div} \lambda = 0 && \text{in } \Omega, \\
&\lambda \cdot \bar{\tau}_\beta = -\bar{\tau}_\beta \cdot \nabla \phi + \bar{h} \cdot \bar{\tau}_\beta, \quad \beta = 1, 2, && \text{on } S_1, \\
&\lambda \cdot \bar{n} = 0 && \text{on } S_1, \\
&\lambda_j = -\nabla_j \phi, \quad j = 1, 2 && \text{on } S_2, \\
&\lambda_3 = 0 && \text{on } S_2.
\end{aligned}
\tag{4.20}
$$

Then, in view of (4.18)–(4.20), the function

$$\tilde{h} = \bar{h} - (\lambda + \nabla\phi) \tag{4.21}$$

is a solution to problem (4.14). For solutions to (4.16) we have

$$\|\bar{h}\|_{1,\sigma,\Omega} \leq c\|d_{,x'}\|_{1,\sigma,S_2}, \quad |\bar{h}_{,t}|_{\sigma,\Omega} \leq c|d_{,x't}|_{\sigma,S_2}, \tag{4.22}$$

where $\sigma \in [1,\infty]$. There exists a Green function for (4.19) (see Lemma 2.11) such that

$$\phi(x,t) = \int_\Omega \sum_{i=1}^2 G(x,y)\partial_{y_i}(\bar{\eta}_1 d_{1,y_i} + \bar{\eta}_2 d_{2,y_i})dy$$

$$= -\int_\Omega \sum_{i=1}^2 \nabla_{y_i} G(x,y)(\bar{\eta}_1 d_{1,y_i} + \bar{\eta}_2 d_{2,y_i})dy, \tag{4.23}$$

where the compatibility condition (4.17) is used. Then

$$\nabla_x \phi(x,t) = \int_\Omega \nabla_x \nabla_{y_i} G(x,y)(\bar{\eta}_1 d_{1,y_i} + \bar{\eta}_2 d_{2,y_i})dy$$

and by the properties of the singular integrals (see the book of Stein [St]) we have (see the remark at the end of the proof of Lemma 2.11, below (2.11).

$$\|\nabla\phi\|_{1,\sigma,\Omega} \leq c\|d_{,x'}\|_{1,\sigma,S_2}, \quad |\nabla\phi_{,t}|_{\sigma\Omega} \leq c|d_{,x't}|_{\sigma,S_2}, \tag{4.24}$$

where $\sigma \in (1,\infty)$.

Utilizing the existence of the Green function to problem (4.20) (see Lemma 2.12) we have

$$\lambda_i(x,t) = \int_{S_1} \frac{\partial G_{i\alpha}}{\partial n_{S_1}}(-\bar{\tau}_\alpha \cdot \nabla\phi + \bar{h} \cdot \bar{\tau}_\alpha)dS_1 + \int_{S_2} \frac{\partial G_{ij}}{\partial n_{S_2}}(-\nabla_j\phi)dS_2,$$

where $\partial_{n_{S_i}}$ means the normal exterior derivative to S_i, $i = 1, 2$. Therefore, the estimates hold (see [St])

$$\|\lambda\|_{1,\sigma,\Omega} \leq c(\|\nabla\phi\|_{1,\sigma,\Omega} + \|\bar{h}\|_{1,\sigma,\Omega}) \leq c\|d_{,x'}\|_{1,\sigma,S_2},$$

$$|\lambda_{,t}|_{\sigma,\Omega} \leq (|\nabla\phi_t|_{\sigma,\Omega} + |\bar{h}_{,t}|_{\sigma,\Omega}) \leq c|d_{,x't}|_{\sigma,S_2}, \tag{4.25}$$

where $\sigma \in (1,\infty)$. From (4.22), (4.24), (4.25), and (4.21) estimate (4.15) holds. This concludes the proof.   □

Let us introduce the new function

$$k = h - \tilde{h}. \tag{4.26}$$

Then, k is a solution to the problem

$$
\begin{aligned}
&k_{,t} - \operatorname{div} \mathbb{T}(h, q) = -v \cdot \nabla h - h \cdot \nabla v - \tilde{h}_{,t} + g \equiv G && \text{in } \Omega^T, \\
&\operatorname{div} k = 0 && \text{in } \Omega^T, \\
&\bar{n} \cdot k = 0, \quad \nu \bar{n} \cdot \mathbb{D}(h) \cdot \bar{\tau}_\alpha + \gamma h \cdot \bar{\tau}_\alpha = 0, \quad \alpha = 1, 2, && \text{on } S_1^T, \\
&k_i = 0, \quad i = 1, 2, \quad h_{3,x_3} = \Delta' d && \text{on } S_2^T, \\
&k|_{t=0} = h(0) - \tilde{h}(0) \equiv k(0) && \text{in } \Omega,
\end{aligned}
\tag{4.27}
$$

where $g = f_{,x_3}$, $\Delta' = \partial_{x_1}^2 + \partial_{x_2}^2$, and v is a solution to problem (1.1). Introducing function δ from (3.7) we see that function w defined by (3.7) also is a solution to problem (3.8). Using that

$$v = w + \delta \tag{4.28}$$

problem (4.27) takes the form

$$
\begin{aligned}
&k_{,t} - \operatorname{div} \mathbb{T}(h, q) = -w \cdot \nabla k - k \cdot \nabla v - \delta \cdot \nabla k - v \cdot \nabla \tilde{h} \\
&\qquad\qquad - \tilde{h} \cdot \nabla v - \tilde{h}_{,t} + g \equiv G && \text{in } \Omega^T, \\
&\operatorname{div} k = 0 && \text{in } \Omega^T, \\
&\bar{n} \cdot k = 0, \quad \nu \bar{n} \cdot \mathbb{D}(h) \cdot \bar{\tau}_\alpha + \gamma h \cdot \bar{\tau}_\alpha = 0, \quad \alpha = 1, 2 && \text{on } S_1^T, \\
&k_i = 0, \quad i = 1, 2, \quad h_{3,x_3} = \Delta' d && \text{on } S_2^T, \\
&k|_{t=0} = k(0) && \text{in } \Omega.
\end{aligned}
\tag{4.29}
$$

Projecting $\operatorname{div} k$ on S_2, we see that

$$\operatorname{div} k|_{S_2} = k_{3,x_3}|_{S_2} = 0. \tag{4.30}$$

Lemma 4.4 *Assume that the following quantities*

$$D_1(t) = |d_1|_{3,6,S_2^t}, \quad \mathcal{V}(t) = |\nabla v|_{3,2,\Omega^t},$$

$$\Lambda_1^2(t) = \int_0^t (\|d_{,x'}\|_{1,2,S_2}^2 + \|d_{,t}\|_{1,S_2}^2 + |f_3|_{4/3,S_2}^2 + |g|_{6/5,\Omega}^2) dt'$$

$$+ \mathcal{A}^2 \sup_t \|d_{,x'}\|_{1,3/2,S_2}^2$$

are finite, where $\mathcal{A} = \mathcal{A}(t)$ is introduced in (4.13). Then, any solution k to problem (4.29) satisfies

$$\|k\|_{V(\Omega^t)} \leq \varphi(D_1, \mathcal{V})(\Lambda_1(t) + |k(0)|_{2,\Omega}), \tag{4.31}$$

where φ is an increasing positive function which is described precisely by the r.h.s. of (4.41).

Proof We shall obtain the energy type estimate for solutions to problem (4.29). Multiplying (4.29)$_1$ by k and integrating over Ω yield

$$\frac{1}{2}\frac{d}{dt}|k|_{2,\Omega}^2 - \int_\Omega \operatorname{div} \mathbb{T}(h, q) \cdot k dx = -\int_\Omega w \cdot \nabla k \cdot k dx$$

$$- \int_\Omega k \cdot \nabla v \cdot k dx - \int_\Omega \delta \cdot \nabla k \cdot k dx - \int_\Omega v \cdot \nabla \tilde{h} \cdot k dx \tag{4.32}$$

$$- \int_\Omega \tilde{h} \cdot \nabla v \cdot k dx - \int_\Omega \tilde{h}_{,t} \cdot k dx + \int_\Omega g \cdot k dx \equiv \int_\Omega G \cdot k dx.$$

Now, we examine the particular terms in (4.32). Integrating by parts, the second term on the l.h.s. takes the form

$$- \int_{S_1} \bar{n} \cdot \mathbb{T}(h, q) \cdot k dS_1 - \int_{S_2} \bar{n} \cdot \mathbb{T}(h, q) \cdot k dS_2 + \frac{\nu}{2} \int_\Omega \mathbb{D}(h) \cdot \mathbb{D}(k) dx \equiv I_1 + I_2 + I_3,$$

where

$$I_1 = -\int_{S_1} \bar{n} \cdot \mathbb{T}(h, q) \cdot \bar{\tau}_\alpha k \cdot \bar{\tau}_\alpha dS_1 = \gamma \int_{S_1} h \cdot \bar{\tau}_\alpha k \cdot \bar{\tau}_\alpha dS_1$$

$$= \gamma |k \cdot \bar{\tau}_\alpha|_{2,S_1}^2 + \gamma \int_{S_1} \tilde{h} \cdot \bar{\tau}_\alpha k \cdot \bar{\tau}_\alpha dS_1,$$

$$I_2 = -\int_{S_2} \mathbb{T}_{33}(h, q) k_3 dS_2 = -\int_{S_2} (2\nu h_{3,x_3} - q) k_3 dS_2$$

$$= -\int_{S_2} (2\nu \tilde{h}_{3,x_3} - q) k_3 dS_2 = -2\nu \int_{S_2} \tilde{h}_{3,x_3} k_3 dS_2 + \int_{S_2} q k_3 dS_2.$$

To examine the last integral, we use the third component of (1.1)$_1$ projected on S_2:

$$d_t + v' \cdot \nabla' d + dh_3 - \nu \Delta' d - \nu \tilde{h}_{3,x_3} - f_3 = -q. \tag{4.33}$$

Using this relation in the last term of I_2, we obtain

$$\int_{S_2} q k_3 dS_2 = \int_{S_2} (-d_{,t} + \nu \Delta' d + \nu \tilde{h}_{3,x_3} + f_3) k_3 dS_2 - \int_{S_2} v' d_{x'} k_3 dS_2$$

$$+ \int_{S_2} d k_3^2 dS_2, \qquad (4.34)$$

where we utilized that $\tilde{h}_3|_{S_2} = 0$ (see (4.14)) and we do not distinguish between dependence on $S_2(-a)$ and $S_2(a)$ because it does not have any influence on estimations.

We estimate the first expression in (4.34) by

$$\varepsilon_1 |k_3|_{4,S_2}^2 + c(1/\varepsilon_1)(|d_{,t}|_{4/3,S_2}^2 + |\Delta' d|_{4/3,S_2}^2 + |f_3|_{4/3,S_2}^2),$$

where the relation $\tilde{h}_{,x_3}|_{S_2} = \Delta' d$ is used, the second term by

$$\varepsilon_2 |k_3|_{4,S_2}^2 + c(1/\varepsilon_2)|v'|_{4,S_2}^2 |d_{,x'}|_{2,S_2}^2,$$

and the last one as follows (see the book of Besov et al. [BIN, Ch. 2, Sect. 10])

$$\int_{S_2} d_1 k_3^2 dS_2 \le |d_1|_{3,S_2} |k_3|_{3,S_2}^2 \le (\varepsilon^{1/3} |\nabla k_3|_{2,\Omega}^2 + c\varepsilon^{-5/3} |k_3|_{2,\Omega}^2)|d_1|_{3,S_2}$$

$$\le \varepsilon_3^{1/3} |\nabla k_3|_{2,\Omega}^2 + c\varepsilon_3^{-5/3} |d_1|_{3,S_2}^6 |k_3|_{2,\Omega}^2.$$

Employing the above estimates in (4.32), we obtain

$$\frac{1}{2} \frac{d}{dt} |k|_{2,\Omega}^2 + \frac{\nu}{2} |\mathbb{D}(k)|_{2,\Omega}^2 + \frac{\gamma}{2} |k \cdot \bar{\tau}_\alpha|_{2,S_1}^2 \le \varepsilon \|k\|_{1,\Omega}^2 + \gamma |\tilde{h} \cdot \bar{\tau}_\alpha|_{2,S_1}^2$$

$$+ c\|\tilde{h}\|_{1,\Omega}^2 + c(1/\varepsilon)(|d_{,t}|_{4/3,S_2}^2 + |\Delta' d|_{4/3,S_2}^2 + |f_3|_{4/3,S_2}^2 + |d_1|_{3,S_2}^6 |k|_{2,\Omega}^2)$$

$$+ \left| \int_\Omega G \cdot k dx \right|.$$

Applying the Korn inequality (9.19) and taking ε sufficiently small, we get

$$\frac{d}{dt} |k|_{2,\Omega}^2 + \nu \|k\|_{1,\Omega}^2 + \gamma |k \cdot \bar{\tau}_\alpha|_{2,S_1}^2 \le c|d_1|_{3,S_2}^6 |k|_{2,\Omega}^2$$

$$(4.35)$$

$$+ c(|d_{,t}|_{4/3,S_2}^2 + |\Delta' d|_{4/3,S_2}^2 + |f_3|_{4/3,S_2}^2 + \|\tilde{h}\|_{1,\Omega}^2) + \left| \int_\Omega G \cdot k dx \right|.$$

Finally, we shall examine the last term on the r.h.s. of (4.35). To this end, we use the r.h.s. of (4.32). The first term of the r.h.s. of (4.32) vanishes. We estimate the second term by

$$\int_\Omega k^2 |\nabla v| dx \le \varepsilon_1 |k|_{6,\Omega}^2 + c(1/\varepsilon_1)|\nabla v|_{3,\Omega}^2 |k|_{2,\Omega}^2.$$

To examine the third term on the r.h.s. of (4.32), we use the definition of δ in (3.7) and express it in the form

$$\int_\Omega b \cdot \nabla k \cdot k dx + \int_\Omega \nabla \varphi \cdot \nabla k \cdot k dx \equiv I_1 + I_2.$$

In view of Sect. 3 and estimates for I_1 and I_2 as defined in (3.11), we have

$$|I_1| \leq c\varrho^{1/6}\|\tilde{d}\|_{1,\Omega}\|k\|_{1,\Omega}^2$$

and

$$|I_2| \leq c(\varepsilon_2\varrho^{\mu-2/3}\sup_{x_3}|\tilde{d}|_{3,S_2} + \varrho^{\mu+1/3}\sup_{x_3}|\tilde{d}_{,x_3}|_{3,S_2})\|k\|_{1,\Omega}^2.$$

We estimate the fourth term by

$$\varepsilon_3|k|_{6,\Omega}^2 + c(1/\varepsilon_3)|v|_{6,\Omega}^2|\nabla\tilde{h}|_{3/2,\Omega}^2,$$

the fifth term by

$$\varepsilon_4|k|_{6,\Omega}^2 + c(1/\varepsilon_4)|\nabla v|_{2,\Omega}^2|\tilde{h}|_{3,\Omega}^2,$$

the sixth by

$$\varepsilon_5|k|_{6,\Omega}^2 + c(1/\varepsilon_5)|\tilde{h}_{,t}|_{6/5,\Omega}^2,$$

and, finally, the last one by

$$\varepsilon_6|k|_{6,\Omega}^2 + c(1/\varepsilon_6)|g|_{6/5,\Omega}^2.$$

Employing the aforementioned consideration in (4.35) and assuming that ϱ, $\varepsilon_1 - \varepsilon_6$ are sufficiently small, we obtain

$$\frac{d}{dt}|k|_{2,\Omega}^2 + \nu\|k\|_{1,\Omega}^2 + \gamma|k \cdot \bar{\tau}_\alpha|_{2,S_1}^2 \leq c(|d_1|_{3,S_2}^6 + |\nabla v|_{3,\Omega}^6)|k|_{2,\Omega}^2$$
$$+ c(|d_{,t}|_{4/3,S_2}^2 + |\Delta'd|_{4/3,S_2}^2 + |f_3|_{4/3,S_2}^2 + \|\tilde{h}\|_{1,\Omega}^2 \qquad (4.36)$$
$$+ \|v\|_{1,\Omega}^2\|\tilde{h}\|_{1,3/2,\Omega}^2 + |\tilde{h}_{,t}|_{6/5,\Omega}^2 + |g|_{6/5,\Omega}^2).$$

From (4.36) it follows

$$\frac{d}{dt}(|k|_{2,\Omega}^2\exp[\nu t - c(|d_1|_{3,6,S_2^t}^6 + |\nabla v|_{3,2,\Omega^t}^6)])$$

$$\leq c(|d_{,t}|_{4/3,S_2}^2 + |\Delta'd|_{4/3,S_2}^2 + |f_3|_{4/3,S_2}^2 + \|\tilde{h}\|_{1,\Omega}^2 + \|v\|_{1,\Omega}^2\|\tilde{h}\|_{1,3/2,\Omega}^2$$

$$+|\tilde{h}_{,t}|^2_{6/5,\Omega} + |g|^2_{6/5,\Omega}) \exp[\nu t - c(|d_1|^6_{3,6,S_2^t} + |\nabla v|^2_{3,2,\Omega^t})].$$

Integrating the inequality with respect to time yields

$$|k(t)|^2_{2,\Omega} \le c \exp c(|d_1|^6_{3,6,S_2^t} + |\nabla v|^2_{3,2,\Omega^t}) \Lambda_1^2(t)$$
$$+ \exp[-\nu t + c(|d_1|^6_{3,6,S_2^t} + |\nabla v|^2_{3,2,\Omega^t})]|k(0)|^2_{2,\Omega}, \tag{4.37}$$

where

$$\Lambda_1^2(t) = |d_{,t}|^2_{4/3,2,S_2^t} + |\Delta' d|^2_{4/3,2,S_2^t} + |f_3|^2_{4/3,2,S_2^t} + \|\tilde{h}\|^2_{1,2,\Omega^t}$$
$$+ \mathcal{A}^2 \|\tilde{h}\|^2_{1,3/2,\infty,\Omega^t} + |\tilde{h}_{,t}|^2_{6/5,2,\Omega^t} + |g|^2_{6/5,2,\Omega^t}$$

and estimate (4.13) was used. In view of (4.15) we simplify Λ_1 to the form

$$\Lambda_1^2(t) = \int_0^t (\|d_{,x'}\|^2_{1,2,S_2} + \|d_{,t}\|^2_{1,S_2} + |f_3|^2_{4/3,S_2} + |g|^2_{6/5,\Omega}) dt' \tag{4.38}$$
$$+ \mathcal{A}^2 \sup_t \|d_{,x'}\|^2_{1,3/2,S_2}.$$

Integrating (4.36) with respect to time and applying (4.37) we derive

$$\|k\|^2_{V(\Omega^t)} \le c(|d_1|^6_{3,6,S_2^t} + |\nabla v|^2_{3,2,\Omega^t})[\exp(|d_1|^6_{3,6,S_2^t}$$
$$+ |\nabla v|^2_{3,2,\Omega^t}) \Lambda_1^2(t) + \exp(-\nu t + c(|d_1|^6_{3,6,S_2^t} \tag{4.39}$$
$$+ |\nabla v|^2_{3,2,\Omega^t}))|k(0)|^2_{2,\Omega}] + \Lambda_1^2(t) + |k(0)|^2_{2,\Omega}.$$

Introducing the notation

$$D_1(t) = |d_1|_{3,6,S_2^t}, \quad \mathcal{V}(t) = |\nabla v|_{3,2,\Omega^t}, \tag{4.40}$$

we can express (4.39) in the form

$$\|k\|^2_{V(\Omega^t)} \le c(D_1^6 + \mathcal{V}^2) \exp(D_1^6 + \mathcal{V}^2) \Lambda_1^2(t)$$
$$+ c(D_1^6 + \mathcal{V}^2) \exp[-\nu t + c(D_1^6 + \mathcal{V}^2)]|k(0)|^2_{2,\Omega} + \Lambda_1^2(t) + |k(0)|^2_{2,\Omega}. \tag{4.41}$$

Inequality (4.41) implies (4.31). This concludes the proof.   □

Corollary 4.5 *Since*

$$\|\tilde{h}\|_{V(\Omega^t)} \le c(|d_{,x'}|_{2,\infty,S_2^t} + \|d_{,x'}\|_{1,2,S_2^t}),$$
$$|\tilde{h}(0)|_{2,\Omega} \le c|d_{,x'}(0)|_{2,S_2} \le c|d_{,x'}|_{2,\infty,S_2^t},$$

we obtain from (4.31) that

$$\|h\|_{V(\Omega^t)} \le \varphi(D_1, \mathcal{V})(\Lambda_1(t) + |h(0)|_{2,\Omega}), \tag{4.42}$$

where $\Lambda_1(t)$ is defined by (4.38).

4.2   A Priori Estimates for Vorticity Component χ

Now, we need to examine solutions to problem (4.9) to derive the energy type estimate. For this we need homogeneous Dirichlet boundary conditions on S_1. For this purpose we introduce the functions $\tilde{\chi}$ as a solution to the problem

$$
\begin{aligned}
\tilde{\chi}_{,t} - \nu \Delta \tilde{\chi} &= 0 && \text{in } \Omega^T, \\
\tilde{\chi} &= \chi_* && \text{on } S_1^T, \\
\tilde{\chi}_{,x_3} &= 0 && \text{on } S_2^T, \\
\tilde{\chi}|_{t=0} &= 0 && \text{in } \Omega,
\end{aligned}
\tag{4.43}
$$

where χ_* is described in $(4.9)_2$. To show the existence of solutions to (4.43) we need the following compatibility conditions

$$\chi_{*,x_3} = 0 \quad \text{on } \bar{S}_1 \cap \bar{S}_2. \tag{4.44}$$

To satisfy $(4.44)_1$ we differentiate χ_* with respect to x_3. It is possible because S_1 is the part of the boundary of cylinder Ω which is parallel to the x_3-axis. Moreover, vectors $\bar{n}|_{S_1}$ and $\bar{\tau}_1|_{S_1}$ do not depend on x_3. Therefore, we need to differentiate the components of velocity only. In χ_* only two-components of velocity v_1 and v_2 appear. Differentiating them with respect to x_3, projecting on S_2, and using $(4.6)_5$, we obtain the compatibility condition $(4.44)_1$ in the form

$$
\begin{aligned}
\chi_{*,x_3}|_{\bar{S}_1 \cap \bar{S}_2} = &-\sum_{i,j=1}^{2} \Big[d_{,x_i}(n_{i,x_j}\tau_{1j} + \tau_{1i,x_j}n_j) + \frac{\gamma}{\nu}d_{,x_j}\tau_{1j} \\
&+ d_{,x_i}\tau_{1i}(\tau_{12,x_1} - \tau_{11,x_2}) \Big] = 0.
\end{aligned}
\tag{4.45}
$$

Then, we can introduce the new function $\chi' = \chi - \tilde{\chi}$, which is a solution to the problem

$$\chi'_{,t} + v \cdot \nabla \chi' - h_3 \chi' + h_2 v_{3,x_1} - h_1 v_{3,x_2} - \nu \Delta \chi' = F_3$$

$$- v \cdot \nabla \tilde{\chi} + h_3 \tilde{\chi} \qquad\qquad\qquad \text{in } \Omega^T,$$

$$\chi' = 0 \qquad\qquad\qquad\qquad\qquad\qquad\qquad \text{on } S_1^T, \quad (4.46)$$

$$\chi'_{,x_3} = 0 \qquad\qquad\qquad\qquad\qquad\qquad\qquad \text{on } S_2^T,$$

$$\chi'|_{t=0} = \chi(0) \qquad\qquad\qquad\qquad\qquad\qquad \text{in } \Omega.$$

Lemma 4.6 *Assume that estimate (4.13) holds. Assume that $h_3 \in L_\infty(0, t;$ $L_3(\Omega))$, $F_3 \in L_2(0, t; L_{6/5}(\Omega))$, $\chi(0) \in L_2(\Omega)$, and*

$$\tilde{\chi} \in L_2(0, t; W_2^1(\Omega)), \quad \tilde{\chi} \in L_\infty(0, t; L_3(\Omega)). \qquad (4.47)$$

Then solutions to problem (4.9) satisfy the inequality

$$\|\chi\|_{V(\Omega^t)}^2 \leq c\mathcal{A}^2 (|h_3|_{3,\infty,\Omega^t}^2 + \sup_t |\tilde{\chi}|_{3,\Omega}^2)$$

$$+ c(|F_3|_{6/5,2,\Omega^t}^2 + \|\tilde{\chi}\|_{V(\Omega^t)}^2 + |\chi(0)|_{2,\Omega}^2). \qquad (4.48)$$

Proof First, we examine problem (4.46). Multiply $(4.46)_1$ by χ', integrate over Ω, and use boundary conditions. Then, we have

$$\frac{d}{dt}|\chi'|_{2,\Omega}^2 + \nu|\nabla \chi'|_{2,\Omega}^2 = \int_\Omega h_3 \chi'^2 dx - \int_\Omega (h_2 v_{3,x_1} - h_1 v_{3,x_2})\chi' dx$$

$$+ \int_\Omega F_3 \chi' dx - \int_\Omega v \cdot \nabla \tilde{\chi} \chi' dx + \int_\Omega h_3 \tilde{\chi} \chi' dx. \qquad (4.49)$$

We estimate the first and last terms together. We have

$$\int_\Omega h_3 \chi'^2 dx + \int_\Omega h_3 \chi' \tilde{\chi} dx = \int_\Omega h_3 \chi' \chi dx,$$

which is bounded by

$$\varepsilon_1 |\chi|_{6,\Omega}^2 + c(1/\varepsilon_1)|h_3|_{3,\Omega}^2 |\chi|_{2,\Omega}^2.$$

The second term from the r.h.s. of (4.49) is bounded by

$$\varepsilon_2 |\chi'|_{6,\Omega}^2 + c(1/\varepsilon_2)|h|_{3,\Omega}^2 |v_{3,x'}|_{2,\Omega}^2,$$

and the third one by

$$\varepsilon_3 |\chi'|_{6,\Omega}^2 + c(1/\varepsilon_3)|F_3|_{6/5,\Omega}^2.$$

Integrating by parts in the fourth term yields

$$\int_\Omega v \cdot \nabla(\tilde{\chi}\chi')dx - \int_\Omega v \cdot \nabla\chi'\tilde{\chi}dx = \int_{S_2} d\tilde{\chi}\chi' dS_2 - \int_\Omega v \cdot \nabla\chi'\tilde{\chi}dx \equiv I_1 + I_2,$$

where

$$|I_1| \leq \varepsilon_4 |\chi'|^2_{4,S_2} + c(1/\varepsilon_4)|d|^2_{2,S_2}|\tilde{\chi}|^2_{4,S_2}$$

and

$$|I_2| \leq \varepsilon_5 |\nabla\chi'|^2_{2,\Omega} + c(1/\varepsilon_5)|v|^2_{6,\Omega}|\tilde{\chi}|^2_{3,\Omega}.$$

Using the above estimates in (4.49) and assuming that $\varepsilon_1 - \varepsilon_5$ are sufficiently small we derive the inequality

$$\frac{d}{dt}|\chi'|^2_{2,\Omega} + \nu\|\chi'\|^2_{1,\Omega} \leq c(|h_3|^2_{3,\Omega}|\chi|^2_{2,\Omega} + |h_3|^2_{3,\Omega}|v_{3,x'}|^2_{2,\Omega} \tag{4.50}$$
$$+ |d|^2_{2,S_2}|\tilde{\chi}|^2_{4,S_2} + |v|^2_{6,\Omega}|\tilde{\chi}|^2_{3,\Omega} + |F_3|^2_{6/5,\Omega}).$$

Integrating (4.50) with respect to time and utilizing (4.13) imply

$$\|\chi'\|^2_{V(\Omega^t)} \leq c\mathcal{A}^2\left(\sup_t |h_3|^2_{3,\Omega} + \sup_t |\tilde{\chi}|^2_{3,\Omega}\right)|d|^2_{2,\infty,S_2^t}\int_0^t |\tilde{\chi}|^2_{4,S_2}dt' \tag{4.51}$$
$$+ c|F_3|^2_{6/5,2,\Omega^t} + |\chi(0)|^2_{2,\Omega}.$$

Since $\chi = \chi' + \tilde{\chi}$ and $\|\chi\|_{V(\Omega^t)} \leq \|\chi'\|_{V(\Omega^t)} + \|\tilde{\chi}\|_{V(\Omega^t)}$ we obtain from (4.51) the inequality (4.48). This concludes the proof.   □

Since Ω is bounded, restrictions (4.47) are equivalent to the following

$$\tilde{\chi} \in L_2(0,t;W_2^1(\Omega)) \cap L_\infty(0,t;L_3(\Omega)), \tag{4.52}$$

where $\tilde{\chi}$ is a solution to (4.43).

Lemma 4.7 (see [Z2, Lemma 6.1], [NZ1, Theorem 1.4]) *For solutions to problem (4.43) we have (see Lemma 2.20)*

$$|\tilde{\chi}|_{3,\infty,\Omega^t} \leq |\chi_*|_{3,\infty,S_1^t} \leq c\|v'\|_{5/6,2,\infty,\Omega^t},$$
$$|\tilde{\chi}|_{2,\infty,\Omega^t} \leq |\chi_*|_{2,\infty,S_1^t} \leq c\|v'\|_{1/2,2,\infty,\Omega^t}, \tag{4.53}$$
$$\|\tilde{\chi}\|_{1,2,\Omega^t} \leq c\|\chi_*\|_{W_2^{1/2,1/4}(S_1^t)} \leq c\|v'\|_{W_2^{1/2,1/4}(S_1^t)} \leq c\|v'\|_{W_2^{1,1/2}(\Omega^t)}.$$

Corollary 4.8 *In view of (4.53) inequality (4.48) takes the form*

$$\|\chi\|^2_{V(\Omega^t)} \leq c(1 + \mathcal{A}^2)(|h_3|^2_{3,\infty,\Omega^t} + |F_3|^2_{6/5,2,\Omega^t}$$
$$+ \|v'\|^2_{W^{1,1/2}_2(\Omega^t)} + \|v'\|^2_{5/6,2,\infty,\Omega^t} + |\chi(0)|^2_{2,\Omega}). \tag{4.54}$$

4.3 Relating v and h: rot–div System

Lemma 4.9 *Assume that estimate (4.13) for the weak solution holds (the quantity $\mathcal{A}$ was introduced there). We set*

$$H^2_1(t) = \sup_t |h|^2_{3,\Omega} + \|h\|^2_{1,2,\Omega^t}. \tag{4.55}$$

We require that $H_1(t)$ and $D_1(t)$, $\mathcal{V}(t)$, $\Lambda_1(t)$ defined in the assumptions of Lemma 4.4 are finite. Assume also that $\chi(0) \in L_2(\Omega)$, $h(0) \in L_2(\Omega)$, $F_3 \in L_2(0,t;L_{6/5}(\Omega))$, and $v' \in L_2(\Omega;H^{1/2}(0,t))$. Then the following inequality holds

$$\|v'\|^2_{V^1(\Omega^t)} \leq \varphi(D_1, \mathcal{V})(\Lambda^2_1(t) + |h(0)|^2_{2,\Omega})$$
$$+ c(1 + \mathcal{A}^2)(H^2_1 + \mathcal{A}^2 + \|v'\|^2_{L_2(\Omega;H^{1/2}(0,t))} \tag{4.56}$$
$$+ |F_3|^2_{6/5,2,\Omega^t} + |\chi(0)|^2_{2,\Omega}).$$

Proof Let Ω' be the cross-section of Ω with the plane perpendicular to the x_3-axis and passing through the point $x_3 \in (-a, a)$. Let S'_1 be the cross-section of S_1 with the same plane. Then S'_1 is the boundary of Ω'. Therefore, the elliptic system (rot, div) reduces to the problem

$$\begin{aligned}
v_{1,x_2} - v_{2,x_1} &= \chi && \text{in } \Omega', \\
v_{1,x_1} + v_{2,x_2} &= -h_3 && \text{in } \Omega', \\
v' \cdot \bar{n}' &= 0 && \text{on } S'_1,
\end{aligned} \tag{4.57}$$

where x_3 is treated as a parameter.

For solution to (4.57) we have

$$\sup_t \|v'\|_{L_2(-a,a;H^1(\Omega'))} + \|v'\|_{L_2(0,t;L_2(-a,a);H^2(\Omega'))}$$
$$\leq c(\|\chi\|_{V(\Omega^t)} + \|h_3\|_{V(\Omega^t)}). \tag{4.58}$$

Moreover, (4.42) implies

$$\|h'\|_{V(\Omega^t)} \leq c\varphi(D_1, \mathcal{V})(\Lambda_1(t) + |h(0)|_{2,\Omega}). \qquad (4.59)$$

Inequalities (4.58) and (4.59) imply

$$\|v'\|^2_{V^1(\Omega^t)} \leq \varphi(D_1, \mathcal{V})(\Lambda_1^2(t) + |h(0)|^2_{2,\Omega})$$
$$+ c(1 + \mathcal{A}^2)(|h_3|^2_{3,\infty,\Omega^t} + \|h_3\|^2_{V(\Omega^t)} + |F_3|^2_{6/5,2,\Omega^t} \quad (4.60)$$
$$+ |\chi(0)|^2_{2,\Omega} + \|v'\|^2_{W_2^{1,1/2}(\Omega^t)} + \|v'\|^2_{5/6,2,\infty,\Omega^t}).$$

Using that

$$\|v'\|_{W_2^{1,1/2}(\Omega^t)} = \|v'\|_{L_2(0,t;H^1(\Omega))} + \|v'\|_{L_2(\Omega;H^{1/2}(0,t))}$$
$$\leq \mathcal{A} + \|v'\|_{L_2(\Omega;H^{1/2}(0,t))}$$

and the interpolation

$$\|v'\|_{5/6,2,\infty,\Omega^t} \leq \varepsilon\|v'\|_{1,2,\infty,\Omega^t} + c(1/\varepsilon)|v'|_{2,\infty,\Omega^t},$$

where the second norm is bounded by $\mathcal{A}$, in the r.h.s. of (4.60), implies (4.56). This concludes the proof. □

Next, we shall obtain an estimate for $\|v\|_{W_{5/3}^{2,1}(\Omega^t)}$ in terms of some norms of h.

Lemma 4.10 *Assume that* $D_1(t) = |d_1|_{3,6,S_2^t}$,

$$\Lambda_1^2(t) = \int_0^t (\|d_{x'}\|^2_{1,2,S_2} + \|d_{,t}\|^2_{1,S_2} + |f_3|^2_{4/3,S_2} + |g|^2_{6/5,\Omega})dt'$$
$$+ \mathcal{A}^2 \sup_t \|d_{,x'}\|^2_{1,3/2,S_2}.$$

Assume that H_1 *is introduced in (4.55) and*

$$H_2 = H_1 + |h|_{10/3,\Omega^t}, \Lambda_2 = \Lambda_1 + |h(0)|_{2,\Omega},$$
$$D_3 = \mathcal{A}(1 + \mathcal{A})H_2 + \varphi(\mathcal{A}) + D_2,$$
$$D_2 = (1 + \mathcal{A})(|F_3|_{6/5,2,\Omega^t} + |\chi(0)|_{2,\Omega}) + |f|_{5/3,\Omega^t}$$
$$+ \|d\|_{W_{5/3}^{7/5,7/10}(S_2^t)} + \|v(0)\|_{W_{5/3}^{4/5}(\Omega)},$$

and $M = c\gamma D_3$ *with* $\gamma > 1$, *where* $\Lambda_2, D_2, D_3,$ *and* H_2 *are finite. Let* Λ_2 *be so small that*

$$\mathcal{A}\varphi(D_1, M)\Lambda_2 \le (\gamma - 1)M, \tag{4.61}$$

where φ already appears in (4.56). Then

$$\|v\|_{W^{2,1}_{5/3}(\Omega^t)} \le M. \tag{4.62}$$

Proof Consider the problem (1.1) written in the form of the Stokes system

$$
\begin{aligned}
&v_{,t} - \operatorname{div} \mathbb{T}(v, p) = -v' \cdot \nabla v - v_3 h + f \equiv \bar{f} && \text{in } \Omega^T, \\
&\operatorname{div} v = 0 && \text{in } \Omega^T, \\
&v \cdot \bar{n} = 0, \quad \nu\bar{n} \cdot \mathbb{D}(v) \cdot \bar{\tau}_\alpha + \gamma v \cdot \bar{\tau}_\alpha = 0, \quad \alpha, 1, 2 && \text{on } S_1^T, \quad (4.63) \\
&v \cdot \bar{n} = d, \quad \bar{n} \cdot \mathbb{D}(v) \cdot \bar{\tau}_\alpha = 0, \quad \alpha = 1, 2 && \text{on } S_2^T, \\
&v|_{t=0} = v(0) && \text{in } \Omega.
\end{aligned}
$$

Applying Lemma 2.21, we have that

$$\|v'\|_{L_{10}(\Omega^T)} \le c\|v'\|_{V^1(\Omega^T)}. \tag{4.64}$$

Then

$$
\begin{aligned}
&|v' \cdot \nabla v|_{5/3, \Omega^T} \le \mathcal{A}\|v'\|_{V^1(\Omega^T)}, \\
&|v_3 h|_{5/3, \Omega^T} \le \mathcal{A}|h|_{10/3, \Omega^T}.
\end{aligned}
\tag{4.65}
$$

In view of the aforementioned estimates and the following estimate for the Stokes system (see [Z8], Theorem A)

$$
\begin{aligned}
\|v\|_{W^{2,1}_\sigma(\Omega^T)} &+ \|\nabla p\|_{L_\sigma(\Omega^T)} \\
&\le c(|\bar{f}|_{\sigma, \Omega^T} + \|d\|_{W^{2-1/\sigma, 1-1/2\sigma}_\sigma(S_2^T)} + \|v(0)\|_{W^{2-2/\sigma}_\sigma(\Omega)}),
\end{aligned}
\tag{4.66}
$$

where $\sigma \in (1, \infty)$, we obtain the following estimate

$$
\begin{aligned}
\|v\|_{W^{2,1}_{5/3}(\Omega^t)} \le c\mathcal{A}(\|v'\|_{V^1(\Omega^t)} + |h|_{10/3, \Omega^t}) \\
+ c(|f|_{5/3, \Omega^t} + \|d\|_{W^{7/5, 7/10}_{5/3}(S_2^t)} + \|v(0)\|_{W^{4/5}_{5/3}(\Omega)}).
\end{aligned}
\tag{4.67}
$$

Taking into account (4.56), it yields

$$
\begin{aligned}
\|v\|_{W^{2,1}_{5/3}(\Omega^t)} \le c\mathcal{A}[\varphi(D_1, \mathcal{V})(\Lambda_1 + |h(0)|_{2,\Omega}) \\
+ (1 + \mathcal{A})(H_1 + \mathcal{A} + \|v'\|_{L_2(\Omega; H^{1/2}(0,t))} + |F_3|_{6/5, 2, \Omega^t} + |\chi(0)|_{2,\Omega})](4.68) \\
+ c\mathcal{A}|h|_{10/3, \Omega^t} + c(|f|_{5/3, \Omega^t} + \|d\|_{W^{7/5, 7/10}_{5/3}(S_2^t)} + \|v(0)\|_{4/5, 5/3, \Omega}).
\end{aligned}
$$

Using the definition of H_2, Λ_2, and D_2, inequality (4.68) can be expressed in the form

$$
\begin{aligned}
\|v\|_{W^{2,1}_{5/3}(\Omega^t)} \le cA[&\varphi(D_1, \mathcal{V})\Lambda_2 + (1 + A)H_2 + \varphi(A) \\
&+ (1 + A)\|v'\|_{L_2(\Omega;H^{1/2}(0,t))}] + cD_2.
\end{aligned}
\tag{4.69}
$$

By the interpolation (see Lemma 2.24)

$$
\|v'\|_{L_2(\Omega;H^{1/2}(0,t))} \le \varepsilon \|v'\|_{W^{2,1}_{5/3}(\Omega^t)} + c(1/\varepsilon)|v'|_{2,\Omega^t},
$$

we obtain from (4.69) the inequality

$$
\|v\|_{W^{2,1}_{5/3}(\Omega^t)} \le cA\varphi(D_1, \mathcal{V})\Lambda_2 + cA(1 + A)H_2 + \varphi(A) + cD_2.
\tag{4.70}
$$

We have $\mathcal{V}(t) = |\nabla v|_{3,2,\Omega^t}$ and the imbedding (see Lemma 2.25)

$$
|\nabla v|_{3,2,\Omega^t} \le c\|v\|_{W^{2,1}_{5/3}(\Omega^t)}.
$$

Introducing the notation $\|v\|_{W^{2,1}_{5/3}(\Omega^t)} = M(t)$ we can express (4.70) in the form

$$
M \le cA\varphi(D_1, M)\Lambda_2 + cD_3,
\tag{4.71}
$$

where D_3 is introduced in the assumptions of this lemma.

Assume that $M = c\gamma D_3$, $\gamma > 1$, and Λ_2 is so small that

$$
A\varphi(D_1, c\gamma D_3)\Lambda_2 \le (\gamma - 1)cD_3.
\tag{4.72}
$$

This implies (4.61) and (4.62), so we end the proof.   □

Remark 4.11 Since all quantities in (4.72) depend on time through the time-integral norms we can satisfy (4.72) or (4.62) for any time assuming that time-integral norms in Λ_2 are sufficiently small. Hence for chosen small Λ_2 we can take T large assuming sufficiently fast decay in the time of integrand functions. Then the bound (4.62) will imply the existence of solutions to problem (1.1) (see [Z8]) for large time with some small data.

To increase the regularity of solutions to (1.1) for given $H_2(t)$ we need the following.

Lemma 4.12 *Let $H_2(t)$, $t \le T$, be finite. Let the assumptions of Lemma 4.10 hold. Let $f \in L_2(\Omega^t)$, $t \le T$, $v(0) \in H^1(\Omega)$. Then any solution to problem (1.1) satisfies the inequality*

$$\|v\|_{W_2^{2,1}(\Omega^t)} + |\nabla p|_{2,\Omega^t} \le \varphi(D_1, D_4 H_2 + D_5)(D_4 H_2 + D_5)\Lambda_2$$

$$+ D_6 H_2^2 + D_7 H_2 + D_8 \equiv \varphi_0(D_1, D_4, D_5, D_6, D_7, D_8, H_2) \qquad (4.73)$$

$$\equiv \varphi_0(H_2), \quad t \le T,$$

where we express φ_0 in more explicit form using that

$$M = D_4 H_2 + D_5, \quad D_4 = c\gamma(1 + \mathcal{A})\mathcal{A},$$

$$D_5 = c\gamma(\varphi(\mathcal{A}) + D_2), \quad D_6 = D_4^2 + (1 + \mathcal{A})D_4,$$

$$D_7 = D_4(\varphi(\mathcal{A}) + D_5 + D_2)D_5 + D_4 D_5 + (1 + \mathcal{A})D_5,$$

$$D_8 = (\varphi(\mathcal{A}) + D_5 + D_2)D_5 + c(|f|_{2,\Omega^t} + \|v(0)\|_{1,\Omega} + \|d\|_{W_2^{2-1/2,1-1/4}(S_2^t)}).$$

Proof We have

$$|v' \cdot \nabla v|_{2,\Omega^t} \le |v'|_{10,\Omega^t} |\nabla v|_{5/2,\Omega^t} \le c\|v'\|_{V^1(\Omega^t)} \|v\|_{W_{5/3}^{2,1}(\Omega^t)} \le c\|v'\|_{V^1(\Omega^t)} M.$$

Applying notation of Lemma 4.10 we obtain from (4.56) the estimate

$$\|v'\|_{V^1(\Omega^t)} \le \varphi(D_1, M)\Lambda_2 + c(1 + \mathcal{A})H_1 + \varphi(\mathcal{A}) + M + D_2.$$

Then

$$|v' \cdot \nabla v|_{2,\Omega^t} \le [\varphi(D_1, M)\Lambda_2 + c(1 + \mathcal{A})H_1 + \varphi(\mathcal{A}) + M + D_2]M.$$

Next, we calculate

$$|v_3 h|_{2,\Omega^t} \le |v_3|_{5,\Omega^t} |h|_{10/3,\Omega^t} \le c\|v\|_{W_{5/3}^{2,1}(\Omega^t)} |h|_{10/3,\Omega^t} \le cMH_2.$$

Hence, solutions to Stokes system (4.63) satisfy

$$\|v\|_{W_2^{2,1}(\Omega^t)} + |\nabla p|_{2,\Omega^t} \le \varphi(D_1, M)M\Lambda_2$$

$$+ (\varphi(\mathcal{A}) + M + D_2)M + c(1 + \mathcal{A})MH_2 + c(|f|_{2,\Omega^t}$$

$$+ \|v(0)\|_{1,\Omega} + \|d\|_{2-1/2,2,S_2^t})$$

$$\equiv \varphi(D_1, D_4 H_2 + D_5)(D_4 H_2 + D_5)\Lambda_2$$

$$+ [D_4^2 + (1 + \mathcal{A})D_4]H_2^2 + [D_4(\varphi(\mathcal{A}) + D_5 + D_2) \qquad (4.74)$$

$$+ D_4 D_5 + (1 + \mathcal{A})D_5]H_2 + (\varphi(\mathcal{A}) + D_5 + D_2)D_5$$

$$+ c(|f|_{2,\Omega^t} + \|v(0)\|_{1,\Omega} + \|d\|_{W_2^{2-1/2,1-1/4}(S_2^t)})$$

$$\equiv \varphi(D_1, D_4 H_2 + D_5)(D_4 H_2 + D_5)\Lambda_2 + D_6 H_2^2 + D_7 H_2 + D_8.$$

This implies (4.73) and concludes the proof.   □

Here we prove a priori estimates for solutions to system (4.6).

Lemma 4.13 *Let the assumptions of Lemmas 4.4 and 4.10 hold. Let* $f \in L_2(\Omega^T)$, $v(0) \in H^1(\Omega)$, $g \in L_\sigma(\Omega^T)$, $h(0) \in W_\sigma^{2-2/\sigma}(\Omega)$, *and* $5/3 \leq \sigma \leq 10/3$. *Let* $\Lambda_2(t)$, $t \leq T$, *be defined in the assumptions of Lemma 4.10.*

Then for sufficiently small Λ_2 *the following estimate holds*

$$\|h\|_{W_\sigma^{2,1}(\Omega^t)} + |\nabla q|_{\sigma,\Omega^t} \leq c\gamma D_9, \quad \gamma > 1, \quad t \leq T, \tag{4.75}$$

where D_9 *is defined by (4.77) and* Λ_2 *is so small that (4.80) holds.*

Proof For solutions to problem (4.6) we have

$$\|h\|_{W_\sigma^{2,1}(\Omega^t)} + |\nabla q|_{\sigma,\Omega^t} \leq c(|v \cdot \nabla h|_{\sigma,\Omega^t} + |h \cdot \nabla v|_{\sigma,\Omega^t}) + cD_9, \tag{4.76}$$

where $\sigma \in (1,\infty)$ and

$$D_9(t) = \sum_{i=1}^{2} \|d_{,x_i}\|_{W_\sigma^{2-1/\sigma,1-1/2\sigma}(S_2^t)} + \|\Delta' d\|_{W_\sigma^{1-1/\sigma,1/2-1/2\sigma}(S_2^t)} \tag{4.77}$$
$$+ |g|_{\sigma,\Omega^t} + \|h(0)\|_{2-2/\sigma,\sigma,\Omega}.$$

Now, we examine the first two norms on the r.h.s. of (4.76). By interpolation from Lemma 2.15 we have

$$|v \cdot \nabla h|_{\sigma,\Omega^t} \leq |v|_{\sigma\lambda_1,\Omega^t} |\nabla h|_{\sigma\lambda_2,\Omega^t}$$
$$\leq |v|_{10,\Omega^t}(\varepsilon_1^{1-\varkappa_1} \|h\|_{W_\sigma^{2,1}(\Omega^t)} + \varepsilon_1^{-\varkappa_1}|h|_{2,\Omega^t}) \equiv I_1,$$

where $1/\lambda_1 + 1/\lambda_2 = 1$ and

$$\varkappa_1 = \left(\frac{5}{\sigma} - \frac{5}{\sigma\lambda_2} + 1\right) = \left(\frac{5}{\sigma\lambda_1} + 1\right) = \frac{3}{4} \quad \text{because} \quad \sigma\lambda_1 = 10.$$

Hence

$$I_1 \leq \varepsilon_2^{1/4} \|h\|_{W_\sigma^{2,1}(\Omega^t)} + c\varepsilon_2^{-3/4}|v|_{10,\Omega^t}^4 |h|_{2,\Omega^t},$$

where we used that $\varepsilon_2^{1/4} = \varepsilon_1^{1/4}|v|_{10,\Omega^t}$. Similarly, by Lemma 2.15, we have

$$|h \cdot \nabla v|_{\sigma,\Omega^t} \leq |h|_{\sigma\lambda_1,\Omega^t} |\nabla v|_{\sigma\lambda_2,\Omega^t}$$
$$\leq |\nabla v|_{10/3,\Omega^t}(\varepsilon_3^{1-\varkappa_2} \|h\|_{W_\sigma^{2,1}(\Omega^t)} + c\varepsilon_3^{-\varkappa_2}|h|_{2,\Omega^t}) \equiv I_2,$$

where

$$\varkappa_2 = \left(\frac{5}{\sigma} - \frac{5}{\sigma\lambda_1}\right)\frac{1}{2} = \frac{5}{2\sigma\lambda_2} = \frac{3}{4} \quad \text{because } \sigma\lambda_2 = \frac{10}{3}.$$

Hence

$$I_2 \le \varepsilon_4^{1/4}\|h\|_{W_\sigma^{2,1}(\Omega^t)} + c\varepsilon_4^{-3/4}|\nabla v|_{10/3,\Omega^t}^4|h|_{2,\Omega^t},$$

where $\varepsilon_4^{1/4} = \varepsilon_3^{1/4}|\nabla v|_{10/3,\Omega^t}$.

Employing the estimates in (4.76), using (4.73) and (4.42) in the form

$$\|h\|_{V(\Omega^t)} \le \varphi(D_1, \varphi_0(H_2))\Lambda_2,$$

we obtain

$$\|h\|_{W_\sigma^{2,1}(\Omega^t)} + |\nabla q|_{\sigma,\Omega^t} \le \varphi_1(H_2)\Lambda_2 + cD_9. \tag{4.78}$$

Employing the imbedding (see Lemma 2.25)

$$H_2(t) \le c\|h\|_{W_\sigma^{2,1}(\Omega^t)} \equiv H, \quad \sigma \ge \frac{5}{3},$$

we derive from (4.78) the inequality for $H_0 = H + |\nabla q|_{\sigma,\Omega^t}$

$$H_0 \le \varphi_1(D_1, \cdots, D_8, H_0)\Lambda_2 + cD_9. \tag{4.79}$$

Setting that

$$H_0 \le c\gamma D_9, \quad \gamma > 1$$

and assuming that Λ_2 is so small that

$$\varphi(D_1, \cdots, D_8, c\gamma D_9)\Lambda_2 \le c(\gamma - 1)D_9 \tag{4.80}$$

estimate (4.75) holds. This ends the proof.   □

Finally, we summarize the results of this section.

Theorem 4.14 *Consider systems (1.1)–(1.4) and (4.6). Assume that*

- *initial data are such that $v(0) \in H^1(\Omega)$, $h(0) \in W_\sigma^{2-2/\sigma}(\Omega)$,*
- *external forces satisfy $f \in L_2(\Omega^t)$, $g \in L_{6/5,2}(\Omega^t) \cap L_\sigma(\Omega^t)$, $F_3 \in L_{6/5,2}(\Omega^t)$, $f_3 \in L_{4/3,2}(S_2^t)$,*

- *flux is such that* $d \in W_2^{1/2,1/4}(S_2^t)$, $d \in L_\infty(0,t;W_{3/2}^2(S_2))$, $d_t \in L_2(0,t;H^1(S_2))$, $\sigma \in [5/3,10/3]$, $d_{x'} \in W_\sigma^{2-1/\sigma,1-1/2\sigma}(S_2^t)$, $d_{x'x'} \in W_\sigma^{1-1/\sigma,1/2-1/2\sigma}(S_2^t)$,

Let the following quantities be finite

$$\mathcal{A}^2(t) = \varphi(\|d\|_{1,3,\infty,S_2^t})(|f|^2_{6/5,2,\Omega^t} + \|d\|^2_{1,3,2,S_2^t} + \|d_t\|^2_{1,6/5,2,S_2^t}) + |v(0)|^2_{2,\Omega},$$

$$\Lambda_2^2(t) = \|d_{,x'}\|^2_{1,2,S_2^t} + \|d_{,t}\|^2_{1,2,S_2^t} + |f_3|^2_{4/3,2,S_2^t} + |g|^2_{6/5,2,\Omega^t}$$

$$+ \mathcal{A}^2\|d_{,x'}\|^2_{1,3/2,\infty,S_2^t} + |h(0)|^2_{2,\Omega},$$

$$D_2(t) = (1 + \mathcal{A}(t))(|F_3|_{6/5,2,\Omega^t} + |\chi(0)|_{2,\Omega}) + |f|_{5/3,\Omega^t} + \|d\|_{W_{5/3}^{7/5,7/10}(S_2^t)}$$

$$+ \|v(0)\|_{W_{5/3}^{4/5}(\Omega)},$$

$$D_1(t) = |d_1|_{3,6,S_2^t},$$

$$D_8'(t) = |f|_{2,\Omega^t} + \|v(0)\|_{1,\Omega} + \|d\|_{W_2^{2-1/2,1-1/4}(S_2^t)},$$

$$D_9(t) = \sum_{i=1}^2 \|d_{,x_i}\|_{W_\sigma^{2-1/\sigma,1-1/2\sigma}(S_2^t)} + \|\Delta'd\|_{W_\sigma^{1-1/\sigma,1/2-1/2\sigma}(S_2^t)}$$

$$+ |g|_{\sigma,\Omega^t} + \|h(0)\|_{2-2/\sigma,\sigma,\Omega},$$

where $5/3 \le \sigma \le 10/3$.

Assume that quantity Λ_2 is so small that inequalities (4.61) and (4.80) hold. Then the following estimates are valid

$$\|h\|_{W_\sigma^{2,1}(\Omega^t)} + |\nabla q|_{\sigma,\Omega^t} \le cD_9, \quad 5/3 \le \sigma \le 10/3 \tag{4.81}$$

and

$$\|v\|_{W_2^{2,1}(\Omega^t)} + |\nabla p|_{2,\Omega^t} \le \varphi(D_1,D_2,D_9,\mathcal{A}) + cD_8'. \tag{4.82}$$

Chapter 5
Global Estimates for Solutions to Problem on (v, p)

Abstract In this chapter we prove the global existence of solutions (v, p) and $(h = v_{,x_3}, q = p_{,x_3})$ by applying the step by step in time argument. From Lemma 5.2 we have

$$\|v(kT)\|_{H^1(\Omega)} \leq Q_1(T) + \exp(-\nu kT)\|v(0)\|_{H^1(\Omega)},$$

and Lemma 5.4 implies

$$\|h(kT)\|_{H^1(\Omega)} \leq Q_2(T) + \exp(-\nu kT)\|h(0)\|_{H^1(\Omega)},$$

where $Q_1(T), Q_2(T)$ depend on data. The above inequalities imply that the initial data for $v(kT), h(kT)$ are bounded by constants independent of k, so in an interval $(kT, (k+1)T)$ we have solvability of v, h in $W_2^{2,1}(\Omega \times (kT, (k+1)T))$. Hence the norms are bounded by quantities independent of k. These inequalities follow from considerations from Chaps. 6 to 9. Smallness of $\Lambda_2(t)$, where

$$\Lambda_2(t) = \int_{kT}^{t} (\|d_{x'}\|_{H^1(S_2)}^2 + \|d_t\|_{H^1(S_2)}^2 + |f_3|_{L_{4/3}(S_2)}^2 + |g|_{L_{6/5}(\Omega)}^2) dt'$$
$$+ A \sup_{t'} \|d_{x'}\|_{W_{3/2}^1(S_2)}^2 + |h(kT)|_{L_2(\Omega)}^2, \quad t \in (kT, (k+1)T),$$

is necessary to repeat the proofs of Chap. 4 in interval $(kT, (k+1)T)$. In this chapter we also derive the precise criteria of smallness for parameter $\Lambda_2(T)$ in terms of data.

In this chapter we derive estimates guaranteeing the existence of global solutions. For this we have to show that $\|v(t)\|_{1,\Omega}$ and $\|h(t)\|_{1,\Omega}$ do not increase with time. In reality our considerations are made step by step in time so we have to show that

© Springer Nature Switzerland AG 2019
J. Rencławowicz, W. M. Zajączkowski, *The Large Flux Problem to the Navier-Stokes Equations*, Advances in Mathematical Fluid Mechanics, https://doi.org/10.1007/978-3-030-32330-1_5

$$\|v((k+1)T)\|_{1,\Omega} \le \|v(kT)\|_{1,\Omega},$$
$$\|h((k+1)T)\|_{1,\Omega} \le \|h(kT)\|_{1,\Omega}, \tag{5.1}$$

where $k \in \mathbb{N}_0$ and T will be chosen later. In other words we show the existence of positive constants C_1 and C_2 such that

$$\|v(t)\|_{1,\Omega} \le C_1,$$
$$\|h(t)\|_{1,\Omega} \le C_2. \tag{5.2}$$

This is an extension of results of Chap. 4 step by step in time. For this, we need T sufficiently large while $\Lambda_2(T)$ remains small enough to fulfill results of Chap. 4.

In order to justify (5.1) it is necessary to show first the differential inequality (5.11) and for this we examine artificial elliptic problem (8.1), where the first equation is identity. Next, we analyze properties of such class of functions that satisfy (5.1), so we exploit (8.8) and (8.18).

To show inequality $(5.1)_2$ we have to consider problem (7.1) as well as the artificial elliptic problems (7.2) and (9.1)—where the first equations are identities. Hence we derive inequalities (7.9) and (9.37) which yield (5.34). Then from inequalities (5.13) and (5.36) we deduce (5.1) for sufficiently large T.

Therefore the role of Chaps. 7–9 is auxiliary to Chap. 4 and this chapter.

First we obtain an estimate for velocity in the interval $[kT, (k+1)T]$ with initial data $v(kT) \in H^1(\Omega)$. Introduce the quantity

$$V = \frac{\nu}{4}|\mathbb{D}(v)|_{2,\Omega}^2 + \frac{\gamma}{2}|v \cdot \bar{\tau}_\alpha|_{2,S_1}^2. \tag{5.3}$$

Lemma 5.1 *Let (v, p) be a solution to (1.1). Let σ be defined in (3.7) and (8.11). Assume that $v \in L_\infty(\Omega)$, $d_{x'} \in L_2(S_2)$, $d \in H^{3/2}(S_2)$, $d_t \in L_2(S_2)$, and $\delta \in H^1(\Omega)$. Then*

$$\frac{d}{dt}V + \nu V \le c(|v|_{\infty,\Omega}^2 + |d_{x'}|_{2,S_2}^8)V + c(\|d\|_{3/2,S_2}^2 + |d_t|_{2,S_2}^2$$
$$+ \|\delta\|_{1,\Omega}^2 + |d_{x'}|_{2,S_2}^{10}). \tag{5.4}$$

Proof Multiplying $(1.1)_1$ by $-\text{div}\,\mathbb{T}(v, p)$ and integrating over Ω yield

$$-\int_\Omega v_t \cdot \text{div}\,\mathbb{T}(v, p)dx + |\text{div}\,\mathbb{T}(v, p)|_{2,\Omega}^2$$
$$= \int_\Omega v \cdot \nabla v \text{div}\,\mathbb{T}(v, p)dx + \int_\Omega f \text{div}\,\mathbb{T}(v, p)dx. \tag{5.5}$$

Applying the Hölder and Young inequalities to the r.h.s. terms of (5.5) gives

$$-\int_\Omega v_t \cdot \operatorname{div} \mathbb{T}(v,p) dx + \frac{1}{2}|\operatorname{div} \mathbb{T}(v,p)|^2_{2,\Omega} \le |v \cdot \nabla v|^2_{2,\Omega} + |f|^2_{2,\Omega}. \qquad (5.6)$$

The first term on the l.h.s. of (5.6) equals

$$-\int_\Omega \operatorname{div}\left(v_t \cdot \mathbb{T}(v,p)\right) dx + \frac{\nu}{2}\int_\Omega \mathbb{D}(v_t) \cdot \mathbb{D}(v) dx \equiv I_1.$$

Expressing v in the local coordinate system in a neighborhood of the boundary

$$v = v_\alpha \bar{\tau}_\alpha + v_n \bar{n}, \quad v_\alpha = v \cdot \bar{\tau}_\alpha, \quad v_n = v \cdot \bar{n}, \quad \alpha = 1,2,$$

the first term in I_1 takes the form

$$-\int_{S_1} n_i v_{jt} \mathbb{T}_{ij}(v,p) dS_1 - \int_{S_2} n_i v_{jt} \mathbb{T}_{ij}(v,p) dS_2$$

$$= -\int_{S_1} n_i (v_{\tau_\alpha,t} \tau_{\alpha j} + v_{n,t} n_j) \mathbb{T}_{ij}(v,p) dS_1$$

$$-\int_{S_2} n_i (v_{\tau_\alpha,t} \tau_{\alpha j} + v_{n,t} n_j) \mathbb{T}_{ij}(v,p) dS_2 \equiv I_2 + I_3,$$

where

$$I_2 = -\int_{S_1} v_{\tau_\alpha,t} n_i \mathbb{T}_{ij} \tau_{\alpha j} dS_1 = \gamma \int_{S_1} v_{\tau_\alpha,t} v_{\tau_\alpha} dS_1 = \frac{\gamma}{2}\frac{d}{dt}|v \cdot \bar{\tau}_\alpha|^2_{2,S_1}$$

We have to emphasize that the summation convention over repeated indices is assumed. Next,

$$I_3 = -\int_{S_2} v_{n,t} n_i n_j \mathbb{T}_{ij}(v,p) dS_2 = -\int_{S_2} d_{,t} \mathbb{T}_{33}(v,p) dS_2.$$

Using quantity (5.3) we obtain from (5.6) the inequality

$$\frac{d}{dt}V + \frac{1}{2}|\operatorname{div} \mathbb{T}(v,p)|^2_{2,\Omega} \le |v \cdot \nabla v|^2_{2,\Omega} + \int_{S_2} d_{,t} \mathbb{T}_{33}(v,p) dS_2 + |f|^2_{2,\Omega}. \qquad (5.7)$$

We estimate the first term on the r.h.s. of (5.7) by

$$|v \cdot \nabla v|^2_{2,\Omega} \le |v|^2_{\infty,\Omega}|\nabla v|^2_{2,\Omega} \equiv I_1.$$

Applying (8.2), we have

$$I_1 \le c|v|^2_{\infty,\Omega}(|\mathbb{D}(v)|^2_{2,\Omega} + |d_{x'}|^2_{2,S_2}) \le c|v|^2_{\infty,\Omega}(V + |d_{x'}|^2_{2,S_2}).$$

In view of the Hölder and Young inequalities the second term on the r.h.s. of (5.7) is estimated by

$$\varepsilon|\mathbb{T}_{33}(v,p)|^2_{2,\Omega} + c(1/\varepsilon)|d_t|^2_{2,S_2} \equiv I_2.$$

By the trace theorem (see Lemma 2.16) we have

$$|\mathbb{T}_{33}(v,p)|^2_{2,S_2} \le c(\|v\|^2_{2,\Omega} + \|p\|^2_{1,\Omega}) \equiv I_3.$$

In view of Lemma 8.4 we have

$$I_3 \le c(|\operatorname{div} \mathbb{T}(v,p)|^2_{2,\Omega} + \|d\|^2_{3/2,S_2} + \|\tilde{d}\|^2_{1,\Omega} + |d_{x'}|^2_{2,S_2})$$
$$\le c(|\operatorname{div} \mathbb{T}(v,p)|^2_{2,\Omega} + \|d\|^2_{3/2,S_2}).$$

Employing the above inequalities in (5.7) and using that ε is sufficiently small, we have

$$\frac{d}{dt}V + \nu|\operatorname{div} \mathbb{T}(v,p)|^2_{2,\Omega} \le c|v|^2_{\infty,\Omega}(V + |d_{x'}|^2_{2,S_2})$$
$$+ c(|d_t|^2_{2,S_2} + \|d\|^2_{3/2,S_2}). \tag{5.8}$$

We examine the following term from the r.h.s. of (5.8),

$$J_1 \equiv |v|^2_{\infty,\Omega}|d_{x'}|^2_{2,S_2}.$$

By interpolation (see [BIN, Sect. 15]) we have

$$J_1 \le c\|v\|^{3/2}_{2,\Omega}|v|^{1/2}_{2,\Omega}|d_{x'}|^2_{2,S_2} \le \varepsilon\|v\|^2_{2,\Omega} + c(1/\varepsilon)|v|^2_{2,\Omega}|d_{x'}|^8_{2,S_2} \equiv J_2.$$

Using (8.2) and (8.31) one obtains

$$J_2 \le \varepsilon(|\operatorname{div} \mathbb{T}(v,p)|^2_{2,\Omega} + \|d\|^2_{3/2,S_2}) + c(1/\varepsilon)(|\mathbb{D}(v)|^2_{2,\Omega} + |d_{x'}|^2_{2,S_2})|d_{x'}|^8_{2,S_2}$$
$$\le \varepsilon(|\operatorname{div} \mathbb{T}(v,p)|^2_{2,\Omega} + \|d\|^2_{3/2,S_2}) + c(1/\varepsilon)(V + |d_{x'}|^2_{2,S_2})|d_{x'}|^8_{2,S_2}.$$

Employing the above estimates in (5.8) and utilizing that ε is sufficiently small yield

$$\frac{d}{dt}V + |\text{div } \mathbb{T}(v, p)|^2_{2,\Omega}$$
$$\leq c(|v|^2_{\infty,\Omega} + |d_{x'}|^8_{2,S_2})V + c(\|d\|^2_{3/2,S_2} + |d_t|^2_{2,S_2} + |d_{x'}|^{10}_{2,S_2}). \tag{5.9}$$

Continuing, (8.8) yields

$$V = |\mathbb{D}(v)|^2_{2,\Omega} + \sum_{\alpha=1}^{2} |v \cdot \bar{\tau}_\alpha|^2_{2,S_1} \leq c\|v\|^2_{1,\Omega}$$
$$\leq c(|\text{div } \mathbb{T}(v, p)|^2_{2,\Omega} + \|\delta\|^2_{1,\Omega} + |d_{x'}|^2_{2,S_2}). \tag{5.10}$$

Using (5.10) in (5.8) implies

$$\frac{d}{dt}V + \nu V \leq c(|v|^2_{\infty,\Omega} + |d_{x'}|^8_{2,S_2})V$$
$$+ c(\|d\|^2_{3/2,S_2} + |d_t|^2_{2,S_2} + \|\delta\|^2_{1,\Omega} + |d_{x'}|^{10}_{2,S_2}). \tag{5.11}$$

This implies (5.4) and concludes the proof.   □

To prove the global existence we need the following result.

Lemma 5.2 *Assume that for V defined in (5.3), $V(0) < \infty$ and the following conditions are fulfilled:*

$$\sup_{k \in \mathbb{N}_0} \int_{kT}^{(k+1)T} (|v(t)|^2_{\infty,\Omega} + |d_{x'}(t)|^8_{2,S_2})dt \leq \nu\frac{T}{2},$$

$$R_1^2(T) = \sup_{k \in \mathbb{N}_0} \int_{kT}^{(k+1)T} (\|d(t)\|^2_{3/2,S_2} + |d_t|^2_{2,S_2}$$
$$+ \|\delta(t)\|^2_{1,\Omega} + |d_{x'}(t)|^{10}_{2,S_2})dt < \infty, \tag{5.12}$$

$$R_2^2(T) = \exp\left[\sup_{k \in \mathbb{N}_0} \int_{kT}^{(k+1)T} \left(|v(t)|^2_{\infty,\Omega} + |d_{x'}(t)|^8_{2,S_2}\right)dt\right] < \infty.$$

Then, for $k \in \mathbb{N}_0$ and $t \in [kT, (k+1)T]$,

$$V(kT) \leq \frac{R_1^2(T)R_2^2(T)}{1 - \exp\left(-\frac{\nu T}{2}\right)} + \exp\left(-\nu\frac{kT}{2}\right)V(0),$$

$$V(t) \leq \frac{R_1^2(T)R_2^2(T)}{1 - \exp\left(-\frac{\nu T}{2}\right)} + \exp\left(-\nu\frac{(k+1)T}{2}\right)V(0). \tag{5.13}$$

Proof From (5.4) we have

$$\frac{d}{dt}\left[V\exp\left(\nu T - \int_{kT}^{t}(|v(t')|^2_{\infty,\Omega} + |d_{x'}|^8_{2,S_2})dt'\right)\right]$$

$$\leq c(\|d\|^2_{3/2,S_2} + |d_t|^2_{2,S_2} + \|\delta\|^2_{1,\Omega} + |d_{x'}|^{10}_{2,S_2}) \qquad (5.14)$$

$$\cdot \exp\left(\nu t - \int_{kT}^{t}(|v(t')|^2_{\infty,\Omega} + |d_{x'}|^8_{2,S_2})dt'\right).$$

Integrating with respect to time from kT to $t \in (kT, (k+1)T]$ yields

$$V(t) \leq c\exp\left(-\nu t + \int_{kT}^{t}(|v(t')|^2_{\infty,\Omega} + |d_{x'}|^8_{2,S_2})dt'\right)$$

$$\cdot \int_{kT}^{t}(\|d(t')\|^2_{3/2,S_2} + |d_t|^2_{2,S_2} + \|\delta(t')\|^2_{1,\Omega} + |d_{x'}(t')|^{10}_{2,S_2})\exp(\nu t')dt' \quad (5.15)$$

$$+ \exp\left[-\nu(t - kT) + \int_{kT}^{t}(|v(t')|^2_{\infty,\Omega} + |d_{x'}|^8_{2,S_2})dt'\right]V(kT).$$

Simplifying (5.15) we get

$$V(t) \leq c\exp\left[\int_{kT}^{t}(|v(t')|^2_{\infty,\Omega} + |d_{x'}|^8_{2,S_2})dt'\right]$$

$$\cdot \int_{kT}^{t}(\|d(t')\|^2_{3/2,S_2} + |d_t|^2_{2,S_2} + \|\delta(t')\|^2_{1,\Omega} + |d_{x'}(t')|^{10}_{2,S_2})dt' \quad (5.16)$$

$$+ \exp\left[-\nu(t - kT) + \int_{kT}^{t}(|v(t')|^2_{\infty,\Omega} + |d_{x'}(t')|^8_{2,S_2})dt'\right]V(kT).$$

Setting $t = (k+1)T$ and assuming that

$$\frac{-\nu T}{2} + \int_{kT}^{(k+1)T}(|v(t)|^2_{\infty,\Omega} + |d_{x'}(t)|^8_{2,S_2})dt \leq 0 \qquad (5.17)$$

and using notation (5.12) we derive

$$V((k+1)T) \leq R_2^2(T)R_1^2(T) + \exp\left(-\frac{\nu T}{2}\right)V(kT). \qquad (5.18)$$

Hence, by iteration, it follows

$$V(kT) \leq \frac{R_2^2(T)R_1^2(T)}{1 - \exp\left(-\frac{\nu T}{2}\right)} + \exp\left(-\nu\frac{kT}{2}\right)V(0). \tag{5.19}$$

Employing (5.19) in (5.16) yields

$$V(t) \leq \frac{R_2^2(T)R_1^2(T)}{1 - \exp\left(-\frac{\nu T}{2}\right)} + \exp\left(-\nu\frac{(k+1)T}{2}\right)V(0), \tag{5.20}$$

where $t \in [kT, (k+1)T]$, $k \in \mathbb{N}_0$. Then (5.13) holds. This concludes the proof.
□

Now, we examine condition $(5.12)_1$. Our aim is to show that $(5.12)_1$ does not imply any restrictions on the magnitude of d. By interpolation the first term under the integral on the l.h.s. of $(5.12)_1$ is bounded by

$$c\int_{kT}^{(k+1)T} \|v\|_{2,\Omega}^{3/2}|v|_{2,\Omega}^{1/2}dt \leq \sup_t |v(t)|_{2,\Omega}^{1/2}\int_{kT}^{(k+1)T} \|v(t)\|_{2,\Omega}^{3/2}dt$$

$$\leq c\sup_t |v(t)|_{2,\Omega}^{1/2}T^{1/4}\left(\int_{kT}^{(k+1)T} \|v(t)\|_{2,\Omega}^2dt\right)^{3/4}.$$

Now our aim is to show that the quantity

$$I = \int_{kT}^{(k+1)T} \|v(t)\|_{2,\Omega}^2dt$$

remains bounded by the same constant for T as large as we want assuming that norms of d_x, d_t, g, and f_3 are sufficiently small (see comments on Λ_2 in the Remark 4.11).

Lemma 5.3 *Let the assumptions of Theorem 4.14 hold. Then*

$$\|v\|_{(2),2,\Omega^t} \leq \varphi(D_1, \cdots, D_9, \Lambda_2), \quad t \leq T, \tag{5.21}$$

where D_1, D_2, D_3, and Λ_2 are defined in Lemma 4.10, D_6, D_7, and D_8 in Lemma 4.12 and D_9 in (4.77). Then estimate (5.21) holds for any finite time T if Λ_2 is sufficiently small for all $t \leq T$.

Proof From (4.13) we have

$$\|v\|_{V(\Omega^t)}^2 \leq \varphi(\sup_t \|d\|_{1,3,S_2})\int_0^t (|f|_{6/5,\Omega}^2 + \|d\|_{1,3,S_2}^2$$
$$+ \|d_t\|_{1,6/5,S_2}^2)dt' + |v(0)|_{2,\Omega}^2 \equiv \mathcal{A}^2. \tag{5.22}$$

Recall that

$$D_1(t) = |d_1|_{3,6,S_2^t},$$

$$\Lambda_1^2(t) = \int_0^t (\|d_{x'}\|_{1,2,S_2}^2 + \|d_t\|_{1,S_2}^2 + |f_3|_{4/3,S_2}^2 + |g|_{6/5,\Omega}^2) dt'$$

$$+ \mathcal{A}^2 \sup_t \|d_{x'}\|_{1,3/2,S_2}^2,$$

$$\Lambda_2^2(t) = \Lambda_1^2(t) + |h(0)|_{2,\Omega}^2,$$

$$D_2(t) = (1+\mathcal{A})(|F_3|_{6/5,2,\Omega^t} + \|\chi(0)\|_{6/5,2,\Omega}) + |f|_{5/3,\Omega^t} + \|d\|_{W_{5/3}^{7/5,7/10}(S_2^t)}$$

$$+ \|v(0)\|_{4/5,5/3,\Omega}.$$

From (4.70) and the imbedding introduced below (4.70) we obtain

$$\|v\|_{W_{5/3}^{2,1}(\Omega^t)} \le c\mathcal{A}\varphi(D_1, \|v\|_{W_{5/3}^{2,1}(\Omega^t)})\Lambda_2 + D_3, \qquad (5.23)$$

where

$$H_1^2 = \sup_t |h|_{3,\Omega}^2 + \|h\|_{1,2,\Omega^t}^2, \quad H_2^2 = H_1^2 + |h|_{10/3,\Omega^t}^2,$$

$$D_3 = c\mathcal{A}(1+\mathcal{A})H_2 + \varphi(\mathcal{A}) + cD_2 \equiv D_4H_2 + D_5.$$

Assuming that Λ_2 is so small that

$$c\mathcal{A}\varphi(D_1, \gamma D_3)\Lambda_2 + D_3 \le \gamma D_3, \qquad (5.24)$$

we obtain that

$$\|v\|_{W_{5/3}^{2,1}(\Omega^t)} \le \gamma D_3, \quad \gamma > 1. \qquad (5.25)$$

Next, (4.74) yields

$$\|v\|_{(2),2,\Omega^t} + |\nabla p|_{2,\Omega^t} \le \varphi(D_1, D_4H_2 + D_5)(D_4H_2 + D_5)\Lambda_2$$
$$+ D_6H_2^2 + D_7H_2 + D_8, \qquad (5.26)$$

where

$$D_6 = D_4^2 + (1+\mathcal{A})D_4, \quad D_7 = D_4(\varphi(\mathcal{A}) + D_2) + D_4D_5 + (1+\mathcal{A})D_5,$$

$$D_8 = (\varphi(\mathcal{A}) + D_2 + D_5)D_5 + c(|f|_{2,\Omega^t} + \|v(0)\|_{1,\Omega} + \|d\|_{(2-1/2),2,S_2^t}).$$

Introducing $H = \|h\|_{2,\sigma,\Omega^t}$, $H_0 = H + |\nabla q|_{\sigma,\Omega^t}$ we obtain (4.79), where D_9 appeared in (4.78). For Λ_2 so small that

$$\varphi(D_1, \cdots, D_8, \gamma D_9)\Lambda_2 \le (\gamma - 1)D_9, \tag{5.27}$$

we have the bound

$$H_0 \le \gamma D_9, \quad \gamma > 1. \tag{5.28}$$

Since $H_2 \le cH_0$ we obtain from (5.26) the estimate

$$\|v\|_{(2),2,\Omega^t} \le \varphi(D_1, \cdots, D_9, \Lambda_2), \tag{5.29}$$

where smallness of Λ_2 holds for any time if time integrals of $d_{x'}$, d_t, f_3, and g are sufficiently small. This concludes the proof. $\qquad\square$

Next, estimates for H' (see (6.2)) and for $|h_{3,t}|_{2,\Omega}$ (see (7.9)) will be derived. Introduce the quantities

$$D_1^2(t) = c(1 + |d|_{2,S_2}^2 + \|v\|_{2,\Omega}^2)\|d_{x'}\|_{1,\Omega}^2 + c\|d_t\|_{1,S_2}^2,$$

$$D_2^2(t) = c(|v|_{\infty,\Omega}^2 + \|v\|_{1,3,\Omega}^2).$$

Then (6.7) takes the form

$$\frac{d}{dt}H' + \nu H' \le D_1^2 + D_2^2 H' + c|h_{3t}|_{2,S_2}^2. \tag{5.30}$$

To find a bound for the last term on the r.h.s. of (5.30) we need estimates from Chap. 7. Introduce the quantities

$$D_3^2(t) = c(\|d_{x't}\|_{1,S_2}^2 + \|d_t\|_{1,S_2}^2 + |d_{tt}|_{4/3,S_2}^2 + |f_{3t}|_{4/3,S_2}^2 + \|v_t\|_{1,\Omega}^2$$

$$+ |g_t|_{6/5,\Omega}^2) + c(\|v\|_{1,\Omega}^2|d_{x't}|_{2,S_2}^2 + \|v_t\|_{1,\Omega}^2|d_{x'}|_{2,S_2}^2),$$

$$D_4^2(t) = c(|v|_{\infty,\Omega}^2 + |\nabla v|_{3,\Omega}^2),$$

$$D_5^2(t) = c(|d_t|_{2,S_2}^2 + |v_t|_{3,\Omega}^2).$$

Then (7.9) yields the inequality

$$\frac{d}{dt}|h_t|_{2,\Omega}^2 + \nu\|h_t\|_{1,\Omega}^2 \le D_3^2(t) + D_4^2|h_t|_{2,\Omega}^2 + D_5^2\|h\|_{1,\Omega}^2. \tag{5.31}$$

In view of (6.8) we derive

$$\frac{d}{dt}|h_t|^2_{2,\Omega} + \nu\|h_t\|^2_{1,\Omega} \leq D^2_3 + D^2_5\|d_{x'}\|^2_{1,S_2} + D^2_4|h_t|^2_{2,\Omega} + D^2_5 H'. \tag{5.32}$$

Introducing the quantity

$$H'_0 = H' + |h_t|^2_{2,\Omega} \tag{5.33}$$

inequalities (5.30) and (5.32) imply

$$\frac{d}{dt}H'_0 + \nu H'_0 \leq D^2_1 + D^2_3 + D^2_5\|d_{x'}\|^2_{1,S_2} + (D^2_2 + D^2_4 + D^2_5)H'_0 \tag{5.34}$$

$$\equiv D^2_6 + D^2_7 H'_0.$$

Lemma 5.4 *Assume that $H'_0(0) < \infty$ and*

$$\sup_{k \in \mathbb{N}_0} \int_{kT}^{(k+1)T} D^2_7(t)dt \leq \nu\frac{T}{2},$$

$$P^2_1(T) = \sup_{k \in \mathbb{N}_0} \int_{kT}^{(k+1)T} D^2_6(t)dt < \infty, \tag{5.35}$$

$$P^2_2(T) = \exp\left[\sup_{k \in \mathbb{N}_0} \int_{kT}^{(k+1)T} D^2_7(t)dt\right] < \infty.$$

Then

$$H'_0(kT) \leq \frac{P^2_1(T)P^2_2(T)}{1 - \exp\left(-\frac{\nu T}{2}\right)} + \exp\left(-\nu\frac{kT}{2}\right)H'_0(0),$$

$$\tag{5.36}$$

$$H'_0(t) \leq \frac{P^2_1(T)P^2_2(T)}{1 - \exp\left(-\frac{\nu T}{2}\right)} + \exp\left(-\nu\frac{(k+1)T}{2}\right)H'_0(0) \text{ for } t \in (kT, (k+1)T].$$

Proof From (5.34) we have

$$\frac{d}{dt}[H'_0 \exp(\nu t - \int_{kT}^t D^2_7(t')dt')] \leq D^2_6 \exp(\nu t - \int_{kT}^t D^2_7(t')dt'). \tag{5.37}$$

Integrating (5.37) with respect to time from $t = kT$ to $t \in (kT, (k+1)T]$ we have

$$H_0'(t) \leq \exp\left(\int_{kT}^t D_7^2(t')dt'\right) \int_{kT}^t D_6^2(t')dt'$$

$$+ \exp\left(-(t - kT) + \int_{kT}^t D_7^2(t')dt'\right) H_0'(kT). \tag{5.38}$$

Setting $t = (k + 1)T$ in (5.38) and applying iteration we get $(5.36)_1$. Next $(5.36)_2$ follows easily. This concludes the proof.   □

Chapter 6
Global Estimates for Solutions to Problem on (h, q)

Abstract In this chapter we examine the system for (h, q), where $h = v_{,x_3}$, $q = p_{,x_3}$ in $\Omega^{kT} = \Omega \times (kT, (k+1)T)$ for $k \in \mathbb{N}_0$:

$$
\begin{aligned}
h_t - \operatorname{div} \mathbb{T}(h, q) &= -v \cdot \nabla h - h \cdot \nabla v + g && \text{in } \Omega^{kT}, \\
\operatorname{div} h &= 0 && \text{in } \Omega^{kT}, \\
h \cdot \bar{n} &= 0 && \text{on } S_1^{kT}, \\
\nu \bar{n} \cdot \mathbb{D}(h) \cdot \bar{\tau}_\alpha + \gamma h \cdot \bar{\tau}_\alpha &= 0, \quad \alpha = 1, 2 && \text{on } S_1^{kT}, \\
h_i &= -d_{,x_i}, \quad i = 1, 2, && \text{on } S_2^{kT}, \\
h_{3,x_3} &= \Delta' d && \text{on } S_2^{kT}, \\
h|_{t=0} &= h(kT) && \text{in } \Omega.
\end{aligned}
$$

We derive the differential inequality for $H' = \frac{1}{4}|\mathbb{D}(h)|^2_{L_2(\Omega)} + \frac{\gamma}{2}|h \cdot \bar{\tau}_\alpha|^2_{L_2(S_1)}$. We use some additional facts on solutions to problem for (h, q) that are established in Chap. 9: estimates for h in $H^1(\Omega)$ and $H^2(\Omega)$.

Denoting $\Omega^{kT} = \Omega \times (kT, (k+1)T)$ and $S_i^{kT} = S_i \times (kT, (k+1)T), i = 1, 2$, we consider problem (4.6) in the form

$$
\begin{aligned}
h_t - \operatorname{div} \mathbb{T}(h, q) &= -v \cdot \nabla h - h \cdot \nabla v + g && \text{in } \Omega^{kT}, \\
\operatorname{div} h &= 0 && \text{in } \Omega^{kT}, \\
h \cdot \bar{n} &= 0 && \text{on } S_1^{kT}, \\
\nu \bar{n} \cdot \mathbb{D}(h) \cdot \bar{\tau}_\alpha + \gamma h \cdot \bar{\tau}_\alpha &= 0, \quad \alpha = 1, 2 && \text{on } S_1^{kT}, \\
h_i &= -d_{,x_i}, \quad i = 1, 2, && \text{on } S_2^{kT}, \\
h_{3,x_3} &= \Delta' d && \text{on } S_2^{kT}, \\
h|_{t=0} &= h(kT) && \text{in } \Omega,
\end{aligned}
\tag{6.1}
$$

© Springer Nature Switzerland AG 2019
J. Rencławowicz, W. M. Zajączkowski, *The Large Flux Problem to the Navier-Stokes Equations*, Advances in Mathematical Fluid Mechanics,
https://doi.org/10.1007/978-3-030-32330-1_6

where $k \in \mathbb{N}_0$. Introduce the notation

$$H' = \frac{1}{4}|\mathbb{D}(h)|^2_{2,\Omega} + \frac{\gamma}{2}|h \cdot \bar{\tau}_\alpha|^2_{2,S_1}. \tag{6.2}$$

Lemma 6.1 *Assume that* $d \in L_2(S_2), d_t \in H^1(S_2), d_{x'} \in H^1(S_2),$ $h_{3t} \in L_2(S_2), f_3 \in L_2(\Omega), g \in L_2(\Omega), v \in W^1_3(\Omega) \bigcap L_\infty(\Omega).$ *Then*

$$\frac{d}{dt}H' + \nu H' \leq c(|v|^2_{\infty,\Omega} + \|v\|^2_{1,3,\Omega})H'$$
$$+ c(1 + |d|^2_{2,S_2} + |v|^2_{\infty,\Omega} + \|v\|^2_{1,3,\Omega})\|d_{x'}\|^2_{1,S_2} \tag{6.3}$$
$$+ c(|h_{3t}|^2_{2,S_2} + \|d_t\|^2_{1,S_2}) + c(|f_3|^2_{2,\Omega} + |g|^2_{2,\Omega}).$$

Proof Multiplying $(6.1)_1$ by $-\mathrm{div}\,\mathbb{T}(h, q)$ and integrating over Ω yield

$$-\int_\Omega h_t \mathrm{div}\,\mathbb{T}(h, q)dx + |\mathrm{div}\,\mathbb{T}(h, q)|^2_{2,\Omega} \tag{6.4}$$
$$= \int_\Omega (v \cdot \nabla h + h \cdot \nabla v)\mathrm{div}\,\mathbb{T}(h, q)dx + \int_\Omega g\,\mathrm{div}\,\mathbb{T}(h, q)dx.$$

The first integral on the l.h.s. takes the form

$$\frac{1}{4}\frac{d}{dt}|\mathbb{D}(h)|^2_{2,\Omega} + \frac{\gamma}{2}\frac{d}{dt}|h \cdot \bar{\tau}_\alpha|^2_{2,S_1} - \int_{S_2} n_j \mathbb{T}_{ij}(h, q)h_{it}dS_2 \equiv I_1.$$

In view of boundary conditions $(6.1)_{5,6}$ the last term in I_1 equals

$$-\int_{S_2} \mathbb{T}_{3i}(h, q)h_{it}dS_2 = \int_{S_2} \mathbb{T}_{3\alpha}(h, q)d_{x_\alpha t}dS_2 - \int_{S_2} \mathbb{T}_{33}(h, q)h_{3t}dS_2.$$

Employing the above expressions in (6.4) and applying the Hölder and Young inequalities, we have

$$\frac{d}{dt}H' + \nu|\mathrm{div}\,\mathbb{T}(h, q)|^2_{2,\Omega} \leq \varepsilon(|\mathbb{T}_{33}|^2_{2,S_2} + |\mathbb{T}_{3\alpha}|^2_{2,S_2})$$
$$+ c(1/\varepsilon)(|h_{3t}|^2_{2,S_2} + |d_{x't}|^2_{2,S_2}) + c(|v \cdot \nabla h|^2_{2,\Omega} + |h \cdot \nabla v|^2_{2,\Omega}) \tag{6.5}$$
$$+ |g|^2_{2,\Omega}).$$

The first term on the r.h.s. of (6.5) is bounded by

$$\varepsilon(\|h\|^2_{2,\Omega} + \|q\|^2_{1,\Omega}).$$

To absorb it by the second term on the l.h.s. of (6.5) we recall estimate (9.37), where $F = \operatorname{div} \mathbb{T}(h, q)$. Then, for sufficiently small ε, inequality (6.5) yields

$$\frac{d}{dt} H' + \nu \|h\|^2_{2,\Omega} + \|q\|^2_{1,\Omega} \leq c(\|d_{x'}\|^2_{1,S_2} + |d_t|^2_{2,S_2}$$

$$+ \|v'\|^2_{1,\Omega} \|d_{x'}\|^2_{1,S_2} + |f_3|^2_{2,\Omega} + |v_3|^2_{2,S_2} |h_3|^2_{2,\Omega} + |d|^2_{2,S_2} \|d_{x'}\|^2_{1,S_2}) \quad (6.6)$$

$$+ c(|h_{3t}|^2_{2,S_2} + |d_{x't}|^2_{2,S_2} + |v \cdot \nabla h|^2_{2,\Omega} + |h \cdot \nabla v|^2_{2,\Omega} + |g|^2_{2,\Omega}).$$

Simplifying, we express (6.6) in the form

$$\frac{d}{dt} H' + \nu H' \leq c(1 + |d|^2_{2,S_2} + \|v\|^2_{1,\Omega} + |v|^2_{\infty,\Omega} + \|v\|^2_{1,3,\Omega}) \|d_{x'}\|^2_{1,S_2}$$

$$+ c\|d_t\|^2_{1,S_2} + c(|v_3|^2_{2,S_2} + |v|^2_{\infty,\Omega} + \|v\|^2_{1,3,\Omega})H' \quad (6.7)$$

$$+ c|h_{3t}|^2_{2,S_2} + c|f_3|^2_{2,\Omega} + c|g|^2_{2,\Omega},$$

where (9.30) was used as follows

$$\|h\|^2_{1,\Omega} \leq c(E_\Omega(h) + \|d_{x'}\|^2_{1,S_2}). \quad (6.8)$$

This implies (6.3) and concludes the proof. $\qquad\square$

Chapter 7
Estimates for h_t

Abstract We analyze the system for h_t in $\Omega^{kT} = \Omega \times (kT, (k+1)T)$ that we obtain by differentiating problem on $h = v_{,x_3}$. We need the Korn inequality and H^1-estimates for solutions to problem on h_t. To achieve this, we use an auxiliary function $\tilde{h}_t$ as a solution to the problem

$$\operatorname{div} \tilde{h}_t = 0 \qquad \text{in } \Omega,$$
$$\tilde{h}_t \cdot \bar{n} = 0 \qquad \text{on } S_1,$$
$$\tilde{h}_{it} = -d_{tx_i}, \quad i = 1, 2, \quad \tilde{h}_{3t} = 0 \quad \text{on } S_2$$

and also the problem for $k_t = h_t - \tilde{h}_t$. Finally, in Corollary 7.4 we derive the differential inequality of the form

$$\frac{d}{dt} H + \nu H \leq \alpha H + L, \quad \text{where}$$

$$H = \frac{1}{4} |\mathbb{D}(h)|^2_{L_2(\Omega)} + \frac{\gamma}{2} |h \cdot \bar{\tau}_\alpha|^2_{L_2(S_1)} + |h_t|^2_{L_2(\Omega)},$$

α, L depend on data and norms of velocity v, γ is the slip coefficient, and ν is the viscosity coefficient.

From (6.1) we obtain the following problem for h_t, denoting $\Omega^{kT} = \Omega \times (kT, (k+1)T)$ and $S_i^{kT} = S_i \times (kT, (k+1)T), i = 1, 2$,

$$h_{tt} - \operatorname{div} \mathbb{T}(h_t, q_t) = -v_t \cdot \nabla h - v \cdot \nabla h_t$$

$$- h_t \cdot \nabla v - h \cdot \nabla v_t + g_t \qquad \text{in } \Omega^{kT},$$

$$\operatorname{div} h_t = 0 \qquad \text{in } \Omega^{kT}, \qquad (7.1)$$

$$h_t \cdot \bar{n} = 0, \quad \nu \bar{n} \cdot \mathbb{D}(h_t) \cdot \bar{\tau}_\alpha + \gamma h_t \cdot \bar{\tau}_\alpha = 0,$$

$$\alpha = 1, 2 \qquad \text{on } S_1^{kT},$$

© Springer Nature Switzerland AG 2019

J. Rencławowicz, W. M. Zajączkowski, *The Large Flux Problem to the Navier-Stokes Equations*, Advances in Mathematical Fluid Mechanics,
https://doi.org/10.1007/978-3-030-32330-1_7

$$h_{ti} = -d_{x_i t}, \quad i = 1, 2, \quad h_{t3,x_3} = \Delta' d_t \qquad \text{on } S_2^{kT},$$

$$h_t|_{t=0} = h_t(kT) \qquad \qquad \text{in } \Omega.$$

We need the Korn inequality and H^1-estimates for solutions to problem (7.1). For this purpose, similarly as in Chap. 9, we consider the problem

$$
\begin{aligned}
&\text{div } \mathbb{T}(h_t, q_t) = F_t, &&\text{in } \Omega, \\
&\text{div } h_t = 0, &&\text{in } \Omega, \\
&h_t \cdot \bar{n} = 0, \quad \nu \bar{n} \cdot \mathbb{D}(h_t) \cdot \bar{\tau}_\alpha + \gamma h_t \cdot \bar{\tau}_\alpha = 0, \quad \alpha = 1, 2, &&\text{on } S_1, \\
&h_{ti} = -d_{tx_i}, \quad i = 1, 2, \quad h_{3t,x_3} = \Delta' d_t, &&\text{on } S_2,
\end{aligned}
$$
(7.2)

which holds on solutions of (7.1) because $(7.2)_1$ is identity. This means that $F_t \equiv \text{div } \mathbb{T}(h_t, q_t)$.

Next, we repeat the considerations from Chap. 9. Introduce function $\tilde{h}_t$ as a solution to the problem

$$
\begin{aligned}
&\text{div } \tilde{h}_t = 0 &&\text{in } \Omega, \\
&\tilde{h}_t \cdot \bar{n} = 0 &&\text{on } S_1, \\
&\tilde{h}_{it} = -d_{tx_i}, \quad i = 1, 2, \quad \tilde{h}_{3t} = 0 &&\text{on } S_2.
\end{aligned}
$$
(7.3)

Repeating the proof of Lemma 9.1 we have the following.

Lemma 7.1 *Assume that $d_{x't} \in H^{1/2+\alpha}(S_2)$, $\alpha = 0, 1$. Then there exists a solution to problem (7.3) such that $\tilde{h}_t \in H^{1+\alpha}(\Omega)$ and*

$$\|\tilde{h}_t\|_{1+\alpha,\Omega} \le c \|d_{x't}\|_{1/2+\alpha,S_2}.$$
(7.4)

In order to follow (9.17) we introduce the function

$$k_t = h_t - \tilde{h}_t,$$
(7.5)

which satisfies

$$
\begin{aligned}
&\text{div } \mathbb{T}(k_t, q_t) = F_t - \text{div } \mathbb{D}(\tilde{h}_t) &&\text{in } \Omega, \\
&\text{div } k_t = 0 &&\text{in } \Omega, \\
&k_t \cdot \bar{n} = 0, \quad \nu \bar{n} \cdot \mathbb{D}(k_t) \cdot \bar{\tau}_\alpha + \gamma k_t \cdot \bar{\tau}_\alpha \\
&\qquad = -(\nu \bar{n} \cdot \mathbb{D}(\tilde{h}_t) \cdot \bar{\tau}_\alpha + \gamma \tilde{h}_t \cdot \bar{\tau}_\alpha), \quad \alpha = 1, 2 &&\text{on } S_1, \\
&k_{it} = 0, \quad i = 1, 2, \quad k_{3t,x_3} = \Delta' d_t - \tilde{h}_{3,x_3 t} &&\text{on } S_2.
\end{aligned}
$$
(7.6)

Repeating the proof of Lemma 9.2 for the problem (7.6) we obtain the following.

Lemma 7.2 *Assume that $E_\Omega(k_t) = |\mathbb{D}(k_t)|_{2,\Omega}^2$, $\mathrm{div}\, k_t = 0$, $k_t \cdot \bar{n}|_{S_1} = 0$, $k_{it}|_{S_2} = 0$, $i = 1, 2$, $k_{3t,x_3} = \Delta' d_t - \tilde{h}_{3t,x_3} \in L_2(S_2)$, $\tilde{h}_{3t} \in H^1(\Omega)$. Then*

$$\|k_t\|_{1,\Omega}^2 \le c(E_\Omega(k_t) + \|\tilde{h}_{3t}\|_{1,\Omega}^2 + |\Delta' d_t - \tilde{h}_{3t,x_3}|_{2,S_2}^2). \tag{7.7}$$

Using (7.5) in (7.7) implies

$$\|h_t\|_{1,\Omega}^2 \le c(E_\Omega(h_t) + \|\tilde{h}_t\|_{1,\Omega}^2 + |\Delta' d_t - \tilde{h}_{3t,x_3}|_{2,S_2}^2). \tag{7.8}$$

Lemma 7.3 *Assume that $d_t \in H^2(S_2)$, $d_{tt} \in L_{4/3}(S_2)$, $d_{x'} \in L_4(S_2)$, $v_t \in L_2(S_2)$, $v \in L_\infty(\Omega) \cap W_3^1(\Omega)$, $f_{3t} \in L_{4/3}(S_2)$, $g_t \in L_{6/5}(\Omega)$. Then for sufficiently regular solutions to (7.1) the inequality holds*

$$\begin{aligned}
\frac{d}{dt}|h_t|_{2,\Omega}^2 + \nu\|h_t\|_{1,\Omega}^2 &\le c(\|d_{x't}\|_{1,S_2}^2 + |d_{tt}|_{4/3,S_2}^2 + |f_{3t}|_{4/3,S_2}^2 + \|d_t\|_{1,S_2}^2 \\
&\quad + |g_t|_{6/5,\Omega}^2) + c(\|v\|_{1,\Omega}^2|d_{x't}|_{2,S_2}^2 + \|v_t\|_{2,S_2}^2|d_{x'}|_{4,S_2}^2) \\
&\quad + c(|v|_{\infty,\Omega}^2 + |\nabla v|_{3,\Omega}^2)|h_t|_{2,\Omega}^2 + c(|d_t|_{2,S_2}^2 + |v_t|_{3,\Omega}^2)\|h\|_{1,\Omega}^2.
\end{aligned} \tag{7.9}$$

Proof Multiplying $(7.1)_1$ by h_t and integrating over Ω yield

$$\begin{aligned}
\frac{1}{2}\frac{d}{dt}|h_t|_{2,\Omega}^2 &- \int_\Omega \mathrm{div}\, \mathbb{T}(h_t, q_t) \cdot h_t dx = -\int_\Omega v_t \cdot \nabla h \cdot h_t dx \\
&- \int_\Omega v \cdot \nabla h_t \cdot h_t dx - \int_\Omega h_t \cdot \nabla v \cdot h_t dx - \int_\Omega h \cdot \nabla v_t \cdot h_t dx \\
&+ \int_\Omega g_t \cdot h_t dx.
\end{aligned} \tag{7.10}$$

The second term on the l.h.s. of (7.10) equals

$$-\int_\Omega (\mathbb{T}_{ij}(h_t, q_t)h_{tj})_{,x_i} dx + \int_\Omega \mathbb{T}_{ij}(h_t, q_t)h_{tj,x_i} dx \equiv I_1.$$

The second term in I_1 equals

$$\frac{1}{2}\int_\Omega |\mathbb{D}(h_t)|^2 dx,$$

and integration by parts with employing the boundary conditions in the first term in I_1 yields

$$-\int_{S_1} n_i \mathbb{T}_{ij}(h_t, q_t) h_{tj} dS_1 - \int_{S_2} n_i \mathbb{T}_{ij}(h_t, q_t) h_{tj} dS_2$$

$$= -\int_{S_1} \bar{n} \cdot \mathbb{D}(h_t) \cdot \bar{\tau}_\alpha h_t \cdot \bar{\tau}_\alpha dS_1 - \int_{S_2} \mathbb{T}_{3j}(h_t, q_t) h_{tj} dS_2$$

$$= \int_{S_1} |h_t \cdot \bar{\tau}_\alpha|^2 dS_1 - \int_{S_2} \mathbb{T}_{33}(h_t, q_t) h_{t3} dS_2 \equiv I_2.$$

To examine the second integral in I_2 we use the time derivative of the third component of (1.1) projected on S_2,

$$\mathbb{T}_{33}(h_t, q_t) = d_{tt} + v' d_{x't} + v'_t d_{x'} + v_3 h_{3t} + v_{3t} h_3 - f_{3t}.$$

Hence,

$$\left| \int_{S_2} \mathbb{T}_{33}(h_t, q_t) \cdot h_{3t} dS_2 \right| \le \varepsilon |h_{3t}|^2_{4,S_2} + c(1/\varepsilon)(|d_{tt}|^2_{4/3,S_2} + |f_{3t}|^2_{4/3,S_2})$$

$$+ \int_{S_2} v' d_{x't} h_{3t} dS_2 + \int_{S_2} v'_t d_{x'} h_{3t} dS_2 + \int_{S_2} v_3 h^2_{3t} dS_2 + \int_{S_2} v_{3t} h_3 h_{3t} dS_2.$$

Next, we estimate the last four terms on the r.h.s. of the above inequality. The third term is bounded by

$$\varepsilon |h_{3t}|^2_{4,S_2} + c(1/\varepsilon)|v'|^2_{4,S_2}|d_{x't}|^2_{2,S_2} \le \varepsilon \|h_{3t}\|^2_{1,\Omega} + c(1/\varepsilon)\|v'\|^2_{1,\Omega}|d_{x't}|^2_{2,S_2},$$

the fourth by

$$\varepsilon |h_{3t}|^2_{4,S_2} + c(1/\varepsilon)|v'_t|^2_{2,S_2}|d_{x'}|^2_{4,S_2} \le \varepsilon \|h_{3t}\|^2_{1,\Omega} + c(1/\varepsilon)\|v'_t\|^2_{2,S_2}|d_{x'}|^2_{4,S_2},$$

the fifth by

$$|v_3|_{\infty,\Omega}|h_{3t}|^2_{2,S_2} \le c|v_3|_{\infty,\Omega}\|h_{3t}\|_{1,\Omega}|h_{3t}|_{2,\Omega} \le \varepsilon \|h_t\|^2_{1,\Omega} + c(1/\varepsilon)|v_3|^2_{\infty,\Omega}|h_{3t}|^2_{2,\Omega},$$

and finally, the sixth by

$$\varepsilon |h_{3t}|^2_{4,S_2} + c(1/\varepsilon)|v_{3t}|^2_{2,S_2}|h_3|^2_{4,S_2} \le \varepsilon \|h_{3t}\|^2_{1,\Omega} + c(1/\varepsilon)|d_t|^2_{2,S_2}\|h_3\|^2_{1,\Omega}.$$

Employing the above estimates in (7.10), using (7.8), and assuming that ε is sufficiently small, we derive the inequality

$$\frac{d}{dt}|h_t|_{2,\Omega}^2 + \nu\|h_t\|_{1,\Omega}^2 + \gamma\int_{S_1}|h_t\cdot\bar{\tau}_\alpha|^2 dS_1 \le c(\|\tilde{h}_t\|_{1,\Omega}^2$$

$$+ |\Delta' d_t - \tilde{h}_{3t,x_3}|_{2,S_2}^2 + |d_{tt}|_{4/3,S_2}^2 + |f_{3t}|_{4/3,S_2}^2)$$

$$+ c(\|v'\|_{1,\Omega}^2|d_{x't}|_{2,S_2}^2 + |v_t'|_{2,S_2}^2|d_{x'}|_{4,S_2}^2 + |v|_{\infty,\Omega}^2|h_{3t}|_{2,\Omega}^2 \qquad (7.11)$$

$$+ |d_t|_{2,S_2}^2\|h_3\|_{1,\Omega}^2) - \int_\Omega v_t\cdot\nabla h\cdot h_t dx - \int_\Omega v\cdot\nabla h_t\cdot h_t dx$$

$$- \int_\Omega h_t\cdot\nabla v\cdot h_t dx - \int_\Omega h\cdot\nabla v_t\cdot h_t dx + \int_\Omega g_t\cdot h_t dx.$$

Now, we estimate the last five integrals from (7.11). The fifth term from the end of (7.11) is bounded by

$$\varepsilon|h_t|_{6,\Omega}^2 + c(1/\varepsilon)|v_t|_{3,\Omega}^2|\nabla h|_{2,\Omega}^2.$$

The fourth term equals

$$\frac{1}{2}\int_\Omega v\cdot\nabla h_t^2 dx = \frac{1}{2}\int_{S_2} v\cdot\bar{n}h_t^2 dS_2 \equiv I.$$

Hence,

$$|I| \le |v|_{\infty,\Omega}|h_t|_{2,S_2}^2 \le c|v|_{\infty,\Omega}\|h_t\|_{1,\Omega}|h_t|_{2,\Omega}$$

$$\le \varepsilon\|h_t\|_{1,\Omega}^2 + c(1/\varepsilon)|v|_{\infty,\Omega}^2|h_t|_{2,\Omega}^2.$$

We estimate the third term from the end of (7.11) by

$$\varepsilon|h_t|_{6,\Omega}^2 + c(1/\varepsilon)|\nabla v|_{3,\Omega}^2|h_t|_{2,\Omega}^2.$$

Integration by parts in the second term from the end of (7.11) yields

$$\int_{S_2} h\cdot\bar{n}v_t\cdot h_t dS_2 + \int_{S_1} h\cdot\bar{n}v_t\cdot h_t dS_1 - \int_\Omega h\cdot\nabla h_t\cdot v_t dx, \qquad (7.12)$$

where the second integral vanishes because $h\cdot\bar{n}|_{S_1} = 0$ and the first integral is expressed in the form

$$\int_{S_2} h\cdot\bar{n}(-v_{\alpha t}d_{x_\alpha t} - d_t v_{\alpha t,x_\alpha})dS_2 = I.$$

Using fractional derivatives we have

$$|I| \le \varepsilon\|h\|_{1,\Omega}^2 + c(1/\varepsilon)\|v\|_{1,\Omega}^2\|d_t\|_{2,S_2}^2.$$

The third integral in (7.12) is bounded by

$$\varepsilon|\nabla h_t|_{2,\Omega}^2 + c(1/\varepsilon)|v_t|_{3,\Omega}^2|h|_{6,\Omega}^2.$$

Finally, the last term on the r.h.s. of (7.11) is estimated by

$$\varepsilon|h_t|_{6,\Omega}^2 + c(1/\varepsilon)|g_t|_{6/5,\Omega}^2.$$

Employing the above estimates in (7.11), utilizing (7.4), and using that ε is sufficiently small yield

$$\begin{aligned}
\frac{d}{dt}|h_t|_{2,\Omega}^2 + \nu\|h_t\|_{1,\Omega}^2 &\le c(\|d_{x't}\|_{1/2,S_2}^2 + |d_{tt}|_{4/3,S_2}^2 + |f_{3t}|_{4/3,S_2}^2 \\
&\quad + \|d_t\|_{1,S_2}^2 + |g_t|_{6/5,\Omega}^2) + c(\|v\|_{1,\Omega}^2|d_{x't}|_{2,S_2}^2 \\
&\quad + \|v_t\|_{1,\Omega}^2|d_{x'}|_{2,S_2}^2) + c(|v|_{\infty,\Omega}^2|h_t|_{2,\Omega}^2 + |d_t|_{2,S_2}^2\|h\|_{1,\Omega}^2 \\
&\quad + |v_t|_{3,\Omega}^2\|h\|_{1,\Omega}^2 + |\nabla v|_{3,\Omega}^2|h_t|_{2,\Omega}^2).
\end{aligned} \tag{7.13}$$

Simplifying (7.13) implies (7.9). This concludes the proof. □

Corollary 7.4 *Let the assumptions of Lemma 7.3 be satisfied. Then holds the following inequality*

$$\frac{d}{dt}H + \nu H \le \alpha H + L, \tag{7.14}$$

where

$$H = \frac{1}{4}|\mathbb{D}(h)|_{2,\Omega}^2 + \frac{\gamma}{2}|h \cdot \bar{\tau}_\alpha|_{2,S_1}^2 + |h_t|_{2,\Omega}^2,$$

$$\alpha(t) = c(|d_t|_{2,S_2}^2 + |v|_{\infty,\Omega}^2 + \|v\|_{1,3,\Omega}^2 + |v_t|_{3,\Omega}^2),$$

$$\begin{aligned}
\int_0^t L(t')dt' &\le c[(1 + |d|_{2,\infty,S_2^t}^2)\|d_{x'}\|_{1,2,S_2^t}^2 \\
&\quad + (|v|_{\infty,2,\Omega^t}^2 + \|v\|_{1,3,2,\Omega^t}^2 + |v_t|_{2,S_2^t}^2)\|d_{x'}\|_{1,2,\infty,S_2^t}^2 \\
&\quad + (|v|_{\infty,2,\Omega^t}^2 + \|v\|_{1,3,2,\Omega^t}^2)\|d_{x't}\|_{1/2,2,\infty,S_2^t}^2 \\
&\quad + \|d_{tt}\|_{4/3,2,S_2^t}^2 + \|d_{x't}\|_{1/2,2,S_2^t}^2 + |f_3|_{2,\Omega^t}^2 \\
&\quad + |f_{3t}|_{4/3,S_2^t}^2 + |g|_{2,\Omega^t}^2 + |g_t|_{2,\Omega^t}^2].
\end{aligned}$$

Proof Let

$$H = H' + |h_t|_{2,\Omega}^2, \tag{7.15}$$

where

$$H' = \frac{1}{4}|\mathbb{D}(h)|^2_{2,\Omega} + \frac{\gamma}{2}|h \cdot \bar{\tau}_\alpha|^2_{2,S_1}$$

is defined in (6.2) Then (6.3), (7.8), (7.9), and (9.30) imply the differential inequality

$$\frac{d}{dt}H + \nu H \leq c(|d_t|^2_{2,S_2} + |v|^2_{\infty,\Omega} + \|v\|^2_{1,3,\Omega} + |v_t|^2_{3,\Omega})H$$

$$+c(1 + |d_t|^2_{2,S_2} + |d|^2_{2,S_2} + |v|^2_{\infty,\Omega} + \|v\|^2_{1,3,\Omega} + |v_t|^2_{2,S_2})\|d_{x'}\|^2_{1,S_2} \quad (7.16)$$

$$+(|v|^2_{\infty,\Omega} + \|v\|^2_{1,3,\Omega})\|d_{x't}\|^2_{1/2,S_2}$$

$$+c(|d_{tt}|^2_{4/3,S_2} + \|d_{x't}\|^2_{1/2,S_2} + |f_3|^2_{2,\Omega} + |f_{3t}|^2_{4/3,S_2} + |g|^2_{2,\Omega} + |g_t|^2_{2,\Omega}).$$

Therefore we can formulate (7.16) in the form of (7.14) and this concludes the proof.   □

Chapter 8
Auxiliary Results: Estimates for (v, p)

Abstract The aim of this chapter is to derive the estimates and prove the existence for stationary Stokes system for (v, p) with the slip boundary conditions in $H^2(\Omega)$ space for velocity v.

$$-\nu \operatorname{div} \mathbb{D}(v) + \nabla p = f \qquad \text{in } \Omega,$$
$$\operatorname{div} v = 0 \qquad \text{in } \Omega,$$
$$v \cdot \bar{n} = 0 \qquad \text{on } S_1,$$
$$\nu \bar{n} \cdot \mathbb{D}(v) \cdot \bar{\tau}_\alpha + \gamma v \cdot \bar{\tau}_\alpha = 0, \quad \alpha = 1, 2 \qquad \text{on } S_1,$$
$$v \cdot \bar{n} = d \qquad \text{on } S_2,$$
$$\bar{n} \cdot \mathbb{D}(v) \cdot \bar{\tau}_\alpha = 0, \quad \alpha = 1, 2 \qquad \text{on } S_2.$$

The proof is divided into two steps. First using the energy inequality and the Korn inequality we show that $v \in H^1(\Omega)$. Next by applying the partition of unity we increase the regularity of v showing that $v \in H^2(\Omega)$.

Consider the problem

$$-\nu \operatorname{div} \mathbb{D}(v) + \nabla p = f \qquad \text{in } \Omega,$$
$$\operatorname{div} v = 0 \qquad \text{in } \Omega,$$
$$v \cdot \bar{n} = 0 \qquad \text{on } S_1,$$
$$\nu \bar{n} \cdot \mathbb{D}(v) \cdot \bar{\tau}_\alpha + \gamma v \cdot \bar{\tau}_\alpha = 0, \quad \alpha = 1, 2 \qquad \text{on } S_1, \tag{8.1}$$
$$v \cdot \bar{n} = d \qquad \text{on } S_2,$$
$$\bar{n} \cdot \mathbb{D}(v) \cdot \bar{\tau}_\alpha = 0, \quad \alpha = 1, 2 \qquad \text{on } S_2,$$

where we assume that $f \equiv -\nu \operatorname{div} \mathbb{D}(v) + \nabla p$, so the first equation is identity.

© Springer Nature Switzerland AG 2019
J. Renclawowicz, W. M. Zajączkowski, *The Large Flux Problem to the Navier-Stokes Equations*, Advances in Mathematical Fluid Mechanics, https://doi.org/10.1007/978-3-030-32330-1_8

Lemma 8.1 *Assume that $E_\Omega(v) = |\mathbb{D}(v)|^2_{2,\Omega} < \infty$, and that $(8.1)_2$–$(8.1)_6$ hold. Assume also that $d_{,x'} \in L_2(S_2)$. Then*

$$\|v\|^2_{1,\Omega} \leq c(|\mathbb{D}(v)|^2_{2,\Omega} + |d_{,x'}|^2_{2,S_2}). \tag{8.2}$$

Proof For sufficiently smooth v, we examine the integral

$$\int_\Omega |\mathbb{D}(v)|^2 dx = \int_\Omega (v_{i,x_j} + v_{j,x_i})^2 dx = 2\int_\Omega v^2_{i,x_j} dx + 2\int_\Omega v_{i,x_j} v_{j,x_i} dx$$
$$= 2\int_\Omega v^2_{i,x_j} dx + 2\int_S n_i v_{i,x_j} v_j dS. \tag{8.3}$$

The boundary integral yields

$$\int_{S_1} n_i v_{i,x_j} v_j dS_1 = -\int_{S_1} n_{i,x_j} v_i v_j dS_1,$$

$$\int_{S_2} n_i v_{i,x_j} v_j dS_2 = \int_{S_2} (v_{3,x_\alpha} v_\alpha + v_{3,x_3} v_3) dS_2$$

$$= \int_{S_2} (d_{,x_\alpha} v_\alpha - v_{\alpha,x_\alpha} v_3) dS_2 = 2\int_{S_2} d_{,x_\alpha} v_\alpha dS_2 + \int_L \bar{n} \cdot vv_3 dL,$$

where it is used that $\partial S_2 = L$, where L is the edge.

Since $L \subset S_1$ condition $(8.1)_3$ implies that the integral over L disappears. Then (8.3) gives

$$|\nabla v|^2_{2,\Omega} \leq |\mathbb{D}(v)|^2_{2,\Omega} + c|d_{,x_\alpha} v_\alpha|_{2,S_2} + c|v \cdot \bar{\tau}_\alpha|^2_{2,S_1} \tag{8.4}$$

From the Korn lemma, i.e., Lemma 2.22, for any positive constants δ, M and for non-axially symmetric domain Ω it holds

$$|v|^2_{2,\Omega} \leq \delta|\nabla v|^2_{2,\Omega} + M|\mathbb{D}(v)|^2_{2,\Omega}, \tag{8.5}$$

where δ can be chosen as small as we need.

By interpolation (see [BIN, Ch. 3, Sect. 15]) inequality (8.4) implies

$$|\nabla v|^2_{2,\Omega} \leq c(|\mathbb{D}(v)|^2_{2,\Omega} + |d_{,x'}|^2_{2,S_2} + |v|^2_{2,\Omega}), \tag{8.6}$$

where $x' = (x_1, x_2)$. Employing (8.5) with sufficiently small δ yields

$$\|v\|^2_{1,\Omega} \leq c(|\mathbb{D}(v)|^2_{2,\Omega} + |d_{,x'}|^2_{2,S_2}). \tag{8.7}$$

This by the density argument gives (8.2) and concludes the proof.   □

Lemma 8.2 *Consider problem (8.1). Assume that* $f \in L_{6/5}(\Omega)$, $d_{x'} \in L_2(S_2)$, $\delta \in H^1(\Omega)$, *where* δ *is defined in (8.11). Then*

$$\|v\|_{1,\Omega}^2 + |v \cdot \bar{\tau}_\alpha|_{2,S_1}^2 \leq c(\|\delta\|_{1,\Omega}^2 + |f|_{6/5,\Omega}^2 + |d_{x'}|_{2,S_2}^2). \tag{8.8}$$

Proof First we transform problem (8.1) to make the Dirichlet boundary conditions homogeneous. Introduce a function α such that

$$\alpha|_{S(-a)} = d_1, \quad \alpha|_{S(a)} = d_2$$

and the vector

$$b = \alpha \bar{e}_3, \quad \bar{e}_3 = (0, 0, 1).$$

We define a function u so that $u = v - b$ which is a solution to the problem

$$\begin{aligned} \operatorname{div} u &= -\operatorname{div} b = -\alpha_{,x_3} && \text{in } \Omega, \\ u \cdot \bar{n} &= 0 && \text{on } S, \end{aligned} \tag{8.9}$$

so the boundary condition for u is homogeneous. Since u is not divergence free we introduce a function φ as a solution to the Neumann problem

$$\begin{aligned} \Delta \varphi &= -\operatorname{div} b && \text{in } \Omega, \\ \bar{n} \cdot \nabla \varphi &= 0 && \text{on } S, \\ \int_\Omega \varphi \, dx &= 0. \end{aligned} \tag{8.10}$$

Introducing the new function

$$w = u - \nabla \varphi = v - (b + \nabla \varphi) = v - \delta \tag{8.11}$$

we see that

$$\operatorname{div} w = 0, \quad w \cdot \bar{n}|_S = 0 \tag{8.12}$$

because

$$\begin{aligned} \operatorname{div} \delta &= 0, \\ \delta \cdot \bar{n}|_S &= d. \end{aligned} \tag{8.13}$$

Then function w is a solution to the problem

$$-\nu \operatorname{div} \mathbb{D}(w+\delta) + \nabla p = f \qquad\qquad \text{in } \Omega,$$

$$\operatorname{div} w = 0 \qquad\qquad \text{in } \Omega,$$

$$w \cdot \bar{n} = 0 \qquad\qquad \text{on } S, \qquad (8.14)$$

$$\nu \bar{n} \cdot \mathbb{D}(w+\delta) \cdot \bar{\tau}_\alpha + \gamma(w+\delta) \cdot \bar{\tau}_\alpha = 0, \quad \alpha = 1,2, \qquad \text{on } S_1,$$

$$\bar{n} \cdot \mathbb{D}(w+\delta) \cdot \bar{\tau}_\alpha = 0, \quad \alpha = 1,2 \qquad \text{on } S_2.$$

Multiplying $(8.14)_1$ by w, integrating over Ω, and using $(8.14)_2$ and the boundary conditions yield

$$\frac{\nu}{2} \int_\Omega \mathbb{D}(w+\delta) \cdot \mathbb{D}(w) dx + \gamma \int_{S_1} (w \cdot \bar{\tau}_\alpha + \delta \cdot \bar{\tau}_\alpha) w \cdot \bar{\tau}_\alpha dS_1 = \int_\Omega f w dx. \qquad (8.15)$$

Hence, we have

$$\nu |\mathbb{D}(w)|_{2,\Omega}^2 + \gamma |w \cdot \bar{\tau}_\alpha|_{2,S_1}^2 \le c(|\mathbb{D}(\delta)|_{2,\Omega}^2 + |\delta \cdot \bar{\tau}_\alpha|_{2,S_1}^2) + \int_\Omega f \cdot w dx. \qquad (8.16)$$

Transformation (8.11) implies

$$|\mathbb{D}(v)|_{2,\Omega}^2 + |v \cdot \bar{\tau}_\alpha|_{2,S_1}^2 \le c(|\mathbb{D}(\delta)|_{2,\Omega}^2 + |\delta \cdot \bar{\tau}_\alpha|_{2,S_1}^2) + \int_\Omega f \cdot (v-\delta) dx. \qquad (8.17)$$

Then Lemma 8.1 and applications of the Hölder and Young inequalities to the last term on the r.h.s. of (8.17) yield

$$\|v\|_{1,\Omega}^2 + |v \cdot \bar{\tau}_\alpha|_{2,S_1}^2 \le c(|\mathbb{D}(\delta)|_{2,\Omega}^2 + |\delta \cdot \bar{\tau}_\alpha|_{2,S_1}^2$$

$$+|\delta|_{6,\Omega}^2 + |d_{,x'}|_{2,S_2}^2 + |f|_{6/5,\Omega}^2).$$

This implies (8.8) and concludes the proof. $\qquad\qquad\qquad\qquad\qquad\qquad \square$

Lemma 8.3 *Assume that* $f \in L_2(\Omega)$, $v \in H^1(\Omega)$, $p \in L_2(\Omega)$, *and* $d \in H^{3/2}(S_2)$. *Then*

$$\|v\|_{2,\Omega} + \|p\|_{1,\Omega} \le c(|f|_{2,\Omega} + \|v\|_{1,\Omega} + |p|_{2,\Omega} + \|d\|_{3/2,S_2}). \qquad (8.18)$$

Proof To prove the Lemma we use the partition of unity introduced in Chap. 2. In this Lemma the (H^2, H^1) regularity for (v,p) is only shown. The existence can be proved by the regularizer technique and the Fredholm theorem.

Let $k \in \mathfrak{M}$ and let $v^{(k)} = v\zeta^{(k)}$, $p^{(k)} = p\zeta^{(k)}$, $f^{(k)} = f\zeta^{(k)}$. Then problem (8.1) for $v^{(k)}$, $p^{(k)}$ takes the form

$$- \nu \Delta v^{(k)} + \nabla p^{(k)} = f^{(k)} - 2\nabla v \nabla \zeta^{(k)} - v\Delta \zeta^{(k)} + p\nabla \zeta^{(k)},$$

$$\operatorname{div} v^{(k)} = v \cdot \nabla \zeta^{(k)}. \tag{8.19}$$

Hence Lemma 2.13 implies

$$\|v^{(k)}\|_{2,\Omega} + \|p^{(k)}\|_{1,\Omega} \leq c(|f^{(k)}|_{2,\Omega} + \|v\|_{1,\Omega^{(k)}} + |p|_{2,\Omega^{(k)}}), \tag{8.20}$$

where $\Omega^{(k)} = \operatorname{supp} \zeta^{(k)}$.

For $k \in \mathfrak{N}_1$ we choose $\zeta^{(k)} \in S_1$ and introduce the new coordinate system $y = (y_1, y_2, y_3)$ with origin at $\xi^{(k)}$. Hence, we have the transformation $y = Y(x)$.

Assume that S_1 is described locally by the equation $y_3 = F_k(y_1, y_2)$, where $\xi^{(k)} = (0, 0, 0)$ in coordinates y. Introduce the new coordinates

$$z_\alpha = y_\alpha, \quad \alpha = 1, 2, \quad z_3 = y_3 - F(y_1, y_2),$$

and define the transformation by $z = \Phi(y)$. In these coordinates $S_1 \cap \bar{\Omega}_k = \{z : z_3 = 0\}$. Let $\Psi = \Phi \circ Y$. Then $\nabla_x = \Psi_x \nabla_z = \Psi_x^j \partial_{z_j}$ and $\hat{\nabla}_z = \Psi_x|_{x=\Psi_k^{-1}(z)} \cdot \nabla_z$. Let $\hat{u}^{(k)}(z) = u(\Psi_k^{-1}(z))$ and $u^{(k)} = \hat{u}^{(k)}(z)\hat{\zeta}^{(k)}(z)$.

For $k \in \mathfrak{N}_1$ and in view of the above-mentioned notation problem (8.1) takes the form

$$- \nu \nabla_z^2 v^{(k)} + \nabla_z p^{(k)} = -\nu(\nabla_z^2 - \hat{\nabla}_z^2)v^{(k)} + (\nabla_z - \hat{\nabla}_z)p^{(k)}$$

$$+ f^{(k)} - 2\hat{\nabla}\hat{v}^{(k)}\hat{\nabla}\hat{\zeta}^{(k)} - \hat{v}^{(k)}\hat{\nabla}\hat{\zeta}^{(k)} + \hat{p}^{(k)}\hat{\nabla}\hat{\zeta}^{(k)},$$

$$\operatorname{div}_z v^{(k)} = (\operatorname{div}_z - \hat{\operatorname{div}}_z)v^{(k)} + \hat{v}^{(k)}\hat{\nabla}_z\hat{\zeta}^{(k)},$$

$$v_3^{(k)}|_{z_3=0} = (v^{(k)}\bar{e}_3 - v^{(k)}\hat{\bar{n}})|_{z_3=0}, \tag{8.21}$$

$$\nu v_{\alpha,z_3}^{(k)}|_{z_3=0} \equiv \nu \bar{e}_3 \cdot \mathbb{D}(v^{(k)}) \cdot \bar{e}_\alpha|_{z_3=0} - \nu v_{3,\tau_\alpha}^{(k)}|_{z_3=0}$$

$$= \nu \bar{e}_3 \cdot \mathbb{D}(v^{(k)}) \cdot \bar{e}_\alpha - \nu v_{3,\tau_\alpha}^{(k)} - \nu \hat{\bar{n}} \cdot \mathbb{D}(v^{(k)}) \cdot \hat{\bar{\tau}}_\alpha - \gamma \hat{v}^{(k)} \cdot \hat{\bar{\tau}}_\alpha$$

$$+ \nu \hat{n}_i(\hat{v}_i \hat{\nabla}_{z_j}\hat{\zeta}^{(k)} + \hat{v}_j \hat{\nabla}_{z_i}\hat{\zeta}^{(k)})\hat{\tau}_{j\alpha}, \quad \alpha = 1, 2, \quad z_3 = 0.$$

In view of Lemma 2.13 we have

$$\|v^{(k)}\|_{2,\mathbb{R}_+^3} + \|p^{(k)}\|_{1,\mathbb{R}_+^3} \leq c\lambda(\|v^{(k)}\|_{2,\mathbb{R}_+^3} + \|p^{(k)}\|_{1,\mathbb{R}_+^3})$$

$$+ c|f^{(k)}|_{2,\mathbb{R}_+^3} + c(\|\hat{v}\|_{1,\mathbb{R}_+^3 \cap \operatorname{supp} \zeta^{(k)}} + \|\hat{p}\|_{1,\mathbb{R}_+^3 \cap \operatorname{supp} \zeta^{(k)}}), \tag{8.22}$$

where $k \in \mathfrak{N}_1$.

Since S_2 is flat we do not have to pass to variables z in the neighborhood of an interior point of S_2. Therefore, for $k \in \mathfrak{N}_2$, problem (8.1) can be expressed in coordinates y with the origin at $\xi^{(k)}$ which are derived from coordinates x by translation only because the origin of coordinates x is located at the middle of the x_3-axis. Then problem (8.1) has the form

$$-\nu \nabla_y^2 v^{(k)} + \nabla_y p^{(k)} = f^{(k)} - 2\nabla v \nabla \zeta^{(k)} + p \nabla \zeta^{(k)},$$

$$\operatorname{div}_y v^{(k)} = v \cdot \nabla \zeta^{(k)},$$

$$v_3^{(k)}|_{y_3=0} = d^{(k)} \tag{8.23}$$

$$\nu v_{\alpha, y_3}^{(k)}|_{y_3=0} = -\gamma v \cdot \bar{\tau}_\alpha + \nu n_i (v_i \nabla_{y_j} \zeta^{(k)} + v_j \nabla_{y_i} \zeta^{(k)}) \tau_{\alpha j},$$

where $\bar{n}|_{y_3=0} = \bar{e}_3$, $\bar{\tau}_1|_{y_3=0} = \bar{e}_1$, and $\bar{\tau}_2|_{y_3=0} = \bar{e}_2$. Moreover, the transformation between coordinates x and y is expressed by $y = x + a$, where $a = (a_1, a_2, a_3)$ is a constant vector.

In view of Lemma 2.13 from Solonnikov [S1] the following estimate for solutions to (8.23) holds

$$\|v^{(k)}\|_{2, \mathbb{R}_+^3} + \|p^{(k)}\|_{1, \mathbb{R}_+^3} \leq c (|f^{(k)}|_{2, \Omega} + \|v\|_{1, \mathbb{R}_+^3 \cap \operatorname{supp} \zeta^{(k)}}$$

$$+ |p|_{2, \mathbb{R}_+^3 \cap \operatorname{supp} \zeta^{(k)}} + \|d^{(k)}\|_{3/2, \mathbb{R}^2}). \tag{8.24}$$

Let $k \in \mathfrak{N}_3$ and $\xi^{(k)} \in L$. Then we introduce new variables $z = (z_1, z_2, z_3)$ such that neighborhood $\Omega^{(k)}$ of $\xi^{(k)}$ is transformed in such a way that L locally is determined by $z_1 = z_2 = 0$ so it is the z_3-axis. Then

$$\Psi_k(S_1 \cap \Omega^{(k)}) = \{z : z_2 = 0\}$$

and

$$\Psi_k(S_2 \cap \Omega^{(k)}) = \{z : z_1 = 0\}.$$

Therefore problem (8.1) in these variables takes the form

$$-\nu \nabla_z^2 v^{(k)} + \nabla_z p^{(k)} = -\nu (\nabla_z^2 - \hat{\nabla}_z^2) v^{(k)} + (\nabla_z - \hat{\nabla}_z) p^{(k)}$$

$$+ f^{(k)} - 2\hat{\nabla} \hat{v}^{(k)} \hat{\nabla} \hat{\zeta}^{(k)} - \hat{v}^{(k)} \hat{\nabla}^2 \hat{\zeta}^{(k)} + \hat{p}^{(k)} \hat{\nabla} \hat{\zeta}^{(k)} \equiv f_1^{(k)} \tag{8.25}$$

in the dihedral right angle $\mathcal{D}_{\pi/2}$ located between two planes $z_1 = 0$ and $z_2 = 0$,

$$\operatorname{div}_z v^{(k)} = (\operatorname{div}_z - \widehat{\operatorname{div}}_z) v^{(k)} + \hat{v}^{(k)} \hat{\nabla} \hat{\zeta}^{(k)} \equiv g^{(k)} \qquad \text{in } \mathcal{D}_{\pi/2},$$

$$v_1^{(k)}|_{z_1=0} = (v^{(k)} \cdot \bar{e}_1 - v^{(k)} \cdot \hat{n})|_{z_1=0} + d^{(k)} \equiv d_1^{(k)} \qquad \text{on } z_1 = 0,$$

$$v_2^{(k)}|_{z_2=0} = (v^{(k)} \cdot \bar{e}_2 - v^{(k)} \cdot \hat{n})|_{z_2=0} \equiv d_2^{(k)} \qquad \text{on } z_2 = 0,$$

$$v_{\alpha,z_1}^{(k)}|_{z_1=0} = (\bar{e}_1 \mathbb{D}(v^{(k)}) \cdot \bar{e}_\alpha - v_{1,z_\alpha}^{(k)} - \hat{\bar{n}} \cdot \mathbb{D}(v^{(k)}) \cdot \hat{\bar{\tau}}_\alpha)|_{z_1=0}$$
$$+\hat{n}_i(\hat{v}_i \hat{\nabla}_{z_j} \hat{\zeta}^{(k)} + \hat{v}_j \hat{\nabla}_{z_i} \hat{\zeta}^{(k)}) \hat{\tau}_{\alpha j}|_{z_1=0} \equiv h_\alpha^{(k)}, \quad \alpha = 2, 3,$$
$$v_{\alpha,z_2}^{(k)}|_{z_2=0} = [-v_{2,z_\alpha}^{(k)} + (\bar{e}_z \cdot \mathbb{D}(v^{(k)}) \cdot \bar{e}_\alpha - \hat{\bar{n}} \cdot \mathbb{D}(v^{(k)}) \cdot \hat{\bar{\tau}}_\alpha)]$$
$$+\hat{n}_i(\hat{v}_i \hat{\nabla}_{z_j} \hat{\zeta}^{(k)} + \hat{v}_j \hat{\nabla}_{z_i} \hat{\zeta}^{(k)}) \cdot \hat{\tau}_{\alpha j}|_{z_2=0} \equiv l_\alpha^{(k)}, \quad \alpha = 1, 3.$$

Hence, problem (8.25) simplifies to

$$
\begin{aligned}
&- \nu \nabla_z^2 v^{(k)} + \nabla_z p^{(k)} = f_1^{(k)} && \text{in } \mathcal{D}_{\pi/2}, \\
&\text{div}_z v^{(k)} = g_1^{(k)} && \text{in } \mathcal{D}_{\pi/2}, \\
&v_1^{(k)}|_{z_1=0} = d_1^{(k)}, \quad v_2^{(k)}|_{z_2=0} = d_2^{(k)}, && (8.26) \\
&v_{\alpha,z_1}^{(k)}|_{z_1=0} = h_\alpha^{(k)}, \quad \alpha = 2, 3, \\
&v_{\alpha,z_2}^{(k)}|_{z_2=0} = l_\alpha^{(k)}, \quad \alpha = 1, 3.
\end{aligned}
$$

To apply Lemma 2.17 we construct functions $u_j^{(k)}$, $j = 1, 2, 3$, in such a way that

$$u_2^{(k)}|_{z_2=0} = d_2^{(k)}, \quad u_{\alpha,z_2}^{(k)}|_{z_2=0} = l_\alpha^{(k)}, \quad \alpha = 1, 3.$$

Introducing the new functions

$$w = v^{(k)} - u^{(k)}, \quad q = p^{(k)},$$

we see that they are solutions to the problem

$$
\begin{aligned}
&- \nu \nabla_z^2 w + \nabla_z q = f_1^{(k)} + \nu \nabla_z^2 u^{(k)} \equiv f_0, \\
&\text{div}_z w = g_1^{(k)} - \text{div}\, u^{(k)} \equiv g_0, \\
&w_1|_{z_1=0} = d_1^{(k)} - u_1^{(k)}|_{z_1=0} \equiv d_0, && (8.27) \\
&w_{\alpha,z_1}|_{z_1=0} = h_\alpha^{(k)} - u_{\alpha,z_1}^{(k)}|_{z_1=0} \equiv h_\alpha, \quad \alpha = 2, 3, \\
&w_2|_{z_2} = 0, \quad w_{\alpha,z_2}|_{z_2=0} = 0, \quad \alpha = 1, 3.
\end{aligned}
$$

Next we construct a reflection $\tilde{w}$ of w with respect to the plane $z_2 = 0$ such that

$$\tilde{w}_2|_{z_2<0} = -w_2|_{z_2>0}, \qquad \tilde{w}_2|_{z_2>0} = w_2|_{z_2>0},$$
$$\tilde{w}_{\alpha,z_2}|_{z_2<0} = w_{\alpha,z_2}|_{z_2>0}, \qquad \tilde{w}_{\alpha,z_2}|_{z_2>0} = w_{\alpha,z_2}|_{z_2>0}, \quad \alpha = 1, 3.$$

After the reflection the problem (8.27) takes the form

$$
\begin{aligned}
-\nu \nabla_z^2 \tilde{w} + \nabla_z \tilde{q} &= \tilde{f}_0, \\
\operatorname{div} \tilde{w} &= \tilde{g}_0, \\
\tilde{w}_1|_{z_1=0} &= \tilde{d}_0, \\
\tilde{w}_{\alpha,z_1}|_{z_1=0} &= \tilde{h}_\alpha, \quad \alpha = 1, 2.
\end{aligned}
\tag{8.28}
$$

Applying Lemma 2.13 to problem (8.28) and using the properties of the reflection and extension (see Lemma 2.15, 2.16) we obtain

$$
\begin{aligned}
\|v^{(k)}\|_{2,\mathcal{D}_{\pi/2}} + \|p^{(k)}\|_{1,\mathcal{D}_{\pi/2}} &\leq c\lambda(\|v^{(k)}\|_{2,\mathcal{D}_{\pi/2}} + \|p^{(k)}\|_{1,\mathcal{D}_{\pi/2}}) \\
&+ c(|f^{(k)}|_{2,\mathcal{D}_{\pi/2}} + \|d^{(k)}\|_{3/2,\Gamma_1}) \\
&+ c(\|v\|_{1,\mathcal{D}_{\pi/2} \cap \operatorname{supp} \zeta^{(k)}} + |p|_{2,\mathcal{D}_{\pi/2} \cap \operatorname{supp} \zeta^{(k)}}),
\end{aligned}
\tag{8.29}
$$

where $\Gamma_1 = \Psi_k(S_2 \cap \operatorname{supp} \zeta^{(k)})$, $\Gamma_2 = \Psi_k(S_1 \cap \operatorname{supp} \zeta^{(k)})$, and $\Gamma_1 \cup \Gamma_2$ is the boundary of $\mathcal{D}_{\pi/2}$.

Passing in (8.22), (8.29) to variables x, adding them to (8.20), (8.24), and applying the properties of the partition of unity we obtain for sufficiently small λ the inequality

$$
\|v\|_{2,\Omega} + \|p\|_{1,\Omega} \leq c(|f|_{2,\Omega} + \|v\|_{1,\Omega} + |p|_{2,\Omega} + \|d\|_{3/2,S_2}).
\tag{8.30}
$$

This implies (8.18) and concludes the proof   □

Lemma 8.4 *Assume that $f \in L_2(\Omega)$, $d \in H^{3/2}(S_2)$. Then the following estimate holds*

$$
\|v\|_{2,\Omega} + \|p\|_{1,\Omega} \leq c(|f|_{2,\Omega} + \|d\|_{3/2,S_2} + |d_{,x'}|_{2,S_2}).
\tag{8.31}
$$

Proof From the form of δ in (8.13) we have

$$
\|\delta\|_{1,\Omega} \leq c\|d\|_{1/2,S_2} \leq \|d\|_{3/2,S_2}.
\tag{8.32}
$$

Let φ be defined by

$$
\begin{aligned}
\operatorname{div} \varphi &= p \quad \text{in } \Omega, \\
\varphi|_S &= 0.
\end{aligned}
\tag{8.33}
$$

Let $p \in L_2(\Omega)$. Then [KP] implies the existence of $\varphi \in H^1(\Omega)$ satisfying the estimate

$$
\|\varphi\|_{1,\Omega} \leq c|p|_{2,\Omega}.
\tag{8.34}
$$

Multiplying $(8.1)_1$ by φ and integrating over Ω yield

$$|p|_{2,\Omega} \leq c(|f|_{2,\Omega} + \|v\|_{1,\Omega}). \tag{8.35}$$

Using (8.32) in (8.8) gives

$$\|v\|_{1,\Omega} \leq c(|f|_{6/5,\Omega} + \|d\|_{1/2,S_2} + |d_{,x'}|_{2,S_2}). \tag{8.36}$$

Employing (8.35) and (8.36) in (8.18) gives (8.31). This concludes the proof.

$\square$

Chapter 9
Auxiliary Results: Estimates for (h, q)

Abstract In this chapter we attain some more refined estimates for (h, q), where $h = v_{,x_3}, q = p_{,x_3}$ and in particular we obtain inequality of the form

$$\|h\|_{H^2(\Omega)} + \|q\|_{H^1(\Omega)} \le c|\mathrm{div}\, \mathbb{T}(h, q)|_{L_2(\Omega)} + \text{ some norms of data.}$$

To get this we consider the stationary Stokes system for (h, q) such that the slip boundary conditions are determined on S_1 and the Dirichlet-Neumann conditions on S_2.

$$\mathrm{div}\, \mathbb{T}(h, q) = F, \quad \mathrm{div}\, h = 0 \qquad \text{in } \Omega,$$
$$h \cdot \bar{n} = 0, \quad \nu\bar{n} \cdot \mathbb{D}(h) \cdot \bar{\tau}_\alpha + \gamma h \cdot \bar{\tau}_\alpha = 0, \quad \alpha = 1, 2 \quad \text{on } S_1,$$
$$h_i = -d_{,x_i}, \quad i = 1, 2, \quad h_{3,x_3} = \Delta'd \qquad \text{on } S_2.$$

First we derive H^1-estimate for h. Since the Dirichlet-Neumann boundary conditions on S_2 are assumed we need to construct some extensions to make Dirichlet boundary conditions homogeneous. Next by using the partition of unity we increase the regularity of h showing that $h \in H^2(\Omega)$. We have to emphasize that the Dirichlet boundary conditions on S_2 imply that the applied Korn inequality holds also for the axially symmetric domain.

Consider the elliptic problem

$$
\begin{aligned}
\mathrm{div}\, \mathbb{T}(h, q) &= F & &\text{in } \Omega, \\
\mathrm{div}\, h &= 0 & &\text{in } \Omega, \\
h \cdot \bar{n} &= 0 & &\text{on } S_1, \\
\nu\bar{n} \cdot \mathbb{D}(h) \cdot \bar{\tau}_\alpha + \gamma h \cdot \bar{\tau}_\alpha &= 0, \quad \alpha = 1, 2, & &\text{on } S_1, \\
h_i &= -d_{,x_i}, \quad i = 1, 2, & &\text{on } S_2, \\
h_{3,x_3} &= \Delta'd & &\text{on } S_2.
\end{aligned}
\tag{9.1}
$$

© Springer Nature Switzerland AG 2019
J. Rencławowicz, W. M. Zajączkowski, *The Large Flux Problem to the Navier-Stokes Equations*, Advances in Mathematical Fluid Mechanics,
https://doi.org/10.1007/978-3-030-32330-1_9

In this chapter we derive some additional estimates for solutions to problem (6.1). For this purpose we consider (9.1), where (h, q) is also a solution to (6.1). This is possible because F is exactly equal to $\operatorname{div} \mathbb{T}(h, q)$, so $(9.1)_1$ is the identity. The main aim of this chapter is to get the inequality

$$\|h\|_{2,\Omega} + \|q\|_{1,\Omega} \le c|\operatorname{div} \mathbb{T}(h, q)|_{2,\Omega} + \text{ some norms of data.}$$

We are not able to show either the Korn inequality or the energy type estimates for solutions to (9.1) because it contains nonhomogeneous Dirichlet boundary conditions on S_2. To make the conditions homogeneous we construct the function $\tilde{h}$ such that

$$\operatorname{div} \tilde{h} = 0 \qquad\qquad\qquad \text{in } \Omega,$$

$$\tilde{h} \cdot \bar{n} = 0 \qquad\qquad\qquad \text{on } S_1, \qquad\qquad (9.2)$$

$$\tilde{h}_i = -d_{,x_i}, \quad i = 1, 2, \quad \tilde{h}_3 = 0 \quad \text{on } S_2.$$

Lemma 9.1 *Assume that $d_{,x'} \in H^{1/2+\alpha}(S_2)$, $\alpha = 0, 1$. Then there exists a solution to problem (9.2) such that $\tilde{h} \in H^{1+\alpha}(\Omega)$ and*

$$\|\tilde{h}\|_{1+\alpha,\Omega} \le c\|d_{,x'}\|_{1/2+\alpha,S_2}. \qquad\qquad (9.3)$$

Proof We show the existence of the function $\tilde{h}$ in a few steps. First we construct a function $\bar{h}$ by the formula

$$\bar{h}_i|_{S_2} = -d_{,x_i}, \quad i = 1, 2,$$
$$\bar{h}_3|_{S_2} = 0. \qquad\qquad\qquad\qquad\qquad (9.4)$$

Let $d \in H^{1/2}(S_2)$ be given. Then, by the inverse trace theorem (see Lemma 2.17), there exists an extension $\tilde{d}$ on Ω such that

$$\tilde{d}|_{S_2} = d, \qquad\qquad\qquad\qquad (9.5)$$

and

$$\|\tilde{d}\|_{1+\alpha,\Omega} \le c\|d\|_{1/2+\alpha,S_2}. \qquad\qquad (9.6)$$

We construct such extension that $\operatorname{supp} \tilde{d}$ is located in some neighborhoods of $S_2(a_i)$, $i = 1, 2$, $a_1 = -a$, and $a_2 = a$. Then the vector function $\bar{h}$ is defined as

$$\bar{h}_i = -\tilde{d}_{,x_i}, \quad i = 1, 2, \quad \bar{h}_3 = 0. \qquad\qquad (9.7)$$

Moreover, in view of the compatibility condition

$$\sum_{i=1}^{2} n_i|_{S_1} d_{,x_i}|_{\bar{S}_1 \cap \bar{S}_2} = 0, \tag{9.8}$$

we can perform a construction of $\bar{h}_i$, $i = 1, 2$, in such a way that

$$\bar{h}_i n_i|_{S_1} = 0. \tag{9.9}$$

Then $\bar{h}$ is a solution to the problem

$$\begin{aligned}
\operatorname{div} \bar{h} &= -\Delta' \tilde{d} & &\text{in } \Omega, \\
\bar{h} \cdot \bar{\tau}_\alpha = \bar{h} \cdot \bar{\tau}_\alpha, \quad \bar{h} \cdot \bar{n} &= 0, \quad \alpha = 1, 2 & &\text{on } S_1, \\
\bar{h}_i = -d_{,x_i}, \quad \bar{h}_3 &= 0, \quad i = 1, 2 & &\text{on } S_2,
\end{aligned} \tag{9.10}$$

where $\Delta' = \partial_{x_1}^2 + \partial_{x_2}^2$. Now we define a function ϕ such that

$$\Delta\phi = -\Delta' \tilde{d}, \quad \bar{n} \cdot \nabla\phi|_S = 0. \tag{9.11}$$

The above construction is possible under the compatibility condition (9.8). Then the function

$$\hat{h} = \bar{h} - \nabla\phi \tag{9.12}$$

is a solution to the problem

$$\begin{aligned}
\operatorname{div} \hat{h} &= 0 & &\text{in } \Omega, \\
\hat{h} \cdot \bar{\tau}_\alpha = -\bar{\tau}_\alpha \cdot \nabla\phi + \bar{h} \cdot \bar{\tau}_\alpha, \quad \alpha = 1, 2 \ \hat{h} \cdot \bar{n} &= 0 & &\text{on } S_1, \\
\hat{h} = -d_{,x_i} - \nabla_i\phi, \quad i = 1, 2, \quad \hat{h}_3 &= 0 & &\text{on } S_2.
\end{aligned} \tag{9.13}$$

To construct the function $\tilde{h}$ we introduce the functions λ and σ such that

$$\begin{aligned}
-\Delta\lambda + \nabla\sigma &= F & &\text{in } \Omega, \\
\operatorname{div}\lambda &= 0 & &\text{in } \Omega, \\
\lambda \cdot \bar{\tau}_\alpha = -\bar{\tau}_\alpha \cdot \nabla\phi + \bar{h} \cdot \bar{\tau}_\alpha, \quad \alpha = 1, 2, \quad \bar{n} \cdot \lambda &= 0 & &\text{on } S_1, \\
\lambda_i = -\nabla_i\phi, \quad i = 1, 2, \quad \lambda_3 &= 0 & &\text{on } S_2.
\end{aligned} \tag{9.14}$$

Then the function

$$\tilde{h} = \bar{h} - (\lambda + \nabla\phi) \tag{9.15}$$

is a solution to problem (9.2). Moreover (9.6) implies

$$\|\bar{h}\|_{2+\alpha,\Omega} \leq c\|d_{,x'}\|_{3/2+\alpha,S_2},$$
$$\|\nabla\phi\|_{2+\alpha,\Omega} \leq c\|d_{,x'}\|_{3/2+\alpha,S_2}, \qquad (9.16)$$
$$\|\lambda\|_{2+\alpha,\Omega} \leq c\|d_{,x'}\|_{3/2+\alpha,S_2}.$$

Hence, (9.3) holds. This concludes the proof.   □

Introduce the function

$$k = h - \tilde{h}. \qquad (9.17)$$

Then k is a solution to the problem

$$\begin{aligned}
\operatorname{div} \mathbb{T}(k, q) &= F - \operatorname{div} \mathbb{D}(\tilde{h}) && \text{in } \Omega, \\
\operatorname{div} k &= 0 && \text{in } \Omega, \\
k \cdot \bar{n} &= 0, \quad \nu\bar{n} \cdot \mathbb{D}(k) \cdot \bar{\tau}_\alpha + \gamma k \cdot \bar{\tau}_\alpha && (9.18) \\
&= -(\nu\bar{n} \cdot \mathbb{D}(\tilde{h}) \cdot \bar{\tau}_\alpha + \gamma\tilde{h} \cdot \bar{\tau}_\alpha), \quad \alpha = 1, 2, && \text{on } S_1, \\
k_i &= 0, \quad i = 1, 2, \quad k_{3,x_3} = \Delta'd - \tilde{h}_{3,x_3} && \text{on } S_2^T.
\end{aligned}$$

Lemma 9.2 *Assume that*

$$E_\Omega(k) = |\mathbb{D}(k)|_{2,\Omega}^2 < \infty, \quad \operatorname{div} k = 0, \quad k \cdot \bar{n}|_{S_1} = 0, \quad k_i|_{S_2} = 0, \quad i = 1, 2,$$
$$k_{3,x_3}|_{S_2} = \Delta'd - \tilde{h}_{3,x_3} \in L_2(S_2), \quad \tilde{h}_3 \in H^1(\Omega).$$

Then

$$\|k\|_{1,\Omega}^2 \leq c(E_\Omega(k) + \|\tilde{h}_3\|_{1,\Omega}^2 + |\Delta'd - \tilde{h}_{3,x_3}|_{2,S_2}^2). \qquad (9.19)$$

Proof We have

$$\int_\Omega |\mathbb{D}(k)|^2 dx = \int_\Omega (k_{i,x_j} + k_{j,x_i})^2 dx = 2\int_\Omega k_{i,x_j}^2 dx + 2\int_\Omega k_{i,x_j} k_{j,x_i} dx. \qquad (9.20)$$

Since k is divergence free we can integrate by parts in the second integral on the r.h.s. of (9.20) to obtain

$$-2\int_{S_1} k_i k_j n_{i,x_j} dS_1 + 2\int_{S_2} (k_{3,x_3} k_3 + k_{3,x_i} k_i) dS_2 \equiv I.$$

Using the boundary conditions on S_2 we have

$$|I| \le c|k'|_{2,S_1}^2 + \varepsilon|k_3|_{2,S_2}^2 + c(1/\varepsilon)|\Delta'd - \tilde{h}_{3,x_3}|_{2,S_2}^2, \tag{9.21}$$

where $k' = (k_1, k_2)$. From (9.20), (9.21), and some interpolation from Lemma 2.15 we get

$$|\nabla k|_{2,\Omega}^2 \le c(E_\Omega(k) + |k'|_{2,\Omega}^2) + \varepsilon|k_3|_{2,S_2}^2 + c(1/\varepsilon)|\Delta'd - \tilde{h}_{3,x_3}|_{2,S_2}^2. \tag{9.22}$$

To estimate the second norm on the r.h.s. of (9.22) we will show that there exist positive constants δ and M such that

$$|k'|_{2,\Omega}^2 \le \delta|\nabla k'|_{2,\Omega}^2 + ME_\Omega(k'), \tag{9.23}$$

δ can be chosen sufficiently small. Moreover, we have

$$k' \cdot \bar{n}|_{S_1} = 0. \tag{9.24}$$

We argue by a contradiction (see similar argument in Solonnikov and Shchadilov [SS]). Assume that such M in (9.23) does not exist. Then for any $m \in \mathbb{N}$ there exists a sequence $k'^{(m)} \in H^1(\Omega)$ such that

$$|k'^{(m)}|_{2,\Omega}^2 \ge \delta|\nabla k'^{(m)}|_{2,\Omega}^2 + mE_\Omega(k'^{(m)}) \equiv G_m(k'^{(m)}).$$

Then for $u^{(m)} = \frac{k'^{(m)}}{|k'^{(m)}|_{2,\Omega}}$ we have

$$|u^{(m)}|_{2,\Omega} = 1, \quad G_m(u^{(m)}) = \frac{G_m(k'^{(m)})}{|k'^{(m)}|_{2,\Omega}^2} \le 1.$$

Therefore, we can choose from the sequence $\{u^{(m)}\}$ a subsequence $\{u^{(m_k)}\}$ which converges weakly in $H^1(\Omega)$ and strongly in $L_2(\Omega)$ to a limit $u \in H^1(\Omega)$. Moreover, $E_\Omega(u) = 0$, so $u = c\eta$, where $\eta = (-x_2, x_1)$.

Since, additionally, $u \cdot \bar{n}|_{S_2} = 0$ we have that $u = 0$. However, this is in contradiction with

$$|u|_{2,\Omega} = \lim_{m_k \to \infty} |u^{(m_k)}|_{2,\Omega} = 1.$$

Hence (9.23) holds. Using it in (9.22) yields

$$|\nabla k|_{2,\Omega}^2 \le cE_\Omega(k) + \varepsilon|k_3|_{2,\Omega}^2 + c(1/\varepsilon)|\Delta d - \tilde{h}_{3,x_3}|_{2,S_2}^2. \tag{9.25}$$

From definition of h_3 we have

$$\int_\Omega h_3 dx = \int_{S_2} dx' \int_{-a}^{a} h_3 dx_3 = \int_{S_2} [v_3(x', a) - v_3(x', -a)] dx'$$

$$= \int_{S_2(a)} d_2 dx' - \int_{S_2(-a)} d_1 dx' = 0,$$

which holds in view of the compatibility conditions. Then by the Poincaré inequality we have

$$|h_3|^2_{2,\Omega} \le c|\nabla h_3|^2_{2,\Omega}. \tag{9.26}$$

Since we need the inequality for k_3 we calculate

$$|k_3|^2_{2,\Omega} = |k_3 + \tilde{h}_3 - \tilde{h}_3|^2_{2,\Omega} \le c|k_3 + \tilde{h}_3|^2_{2,\Omega} + c|\tilde{h}_3|^2_{2,\Omega}$$
$$\le c|\nabla(k_3 + \tilde{h}_3)|^2_{2,\Omega} + c|\tilde{h}_3|^2_{2,\Omega} \le c|\nabla k_3|^2_{2,\Omega} + c\|\tilde{h}_3\|^2_{1,\Omega}. \tag{9.27}$$

Using (9.27) in (9.25) and assuming that ε is sufficiently small we obtain

$$|\nabla k|^2_{2,\Omega} \le c(E_\Omega(k) + \|\tilde{h}_3\|^2_{1,\Omega} + |\Delta'd - \tilde{h}_{3,x_3}|^2_{2,S_2}) \equiv cJ. \tag{9.28}$$

Employing (9.28) in (9.27) gives

$$|k_3|^2_{2,\Omega} \le cJ. \tag{9.29}$$

Hence (9.23), (9.28), and (9.29) imply (9.19). This concludes the proof. □

Employing the relation (9.17) between k and h we obtain from (9.19) the inequality

$$\|h\|^2_{1,\Omega} \le c(E_\Omega(k) + \|\tilde{h}\|^2_{1,\Omega} + |\Delta'd - \tilde{h}_{3,x_3}|^2_{2,S_2})$$
$$\le c(E_\Omega(h) + \|\tilde{h}\|^2_{1,\Omega} + |\Delta'd - \tilde{h}_{3,x_3}|^2_{2,S_2}). \tag{9.30}$$

Next we find an estimate for $E_\Omega(k)$.

Lemma 9.3 (See [RZ6, Z1]) *Assume that* $k \in H^1(\Omega)$, $\tilde{h} \in H^1(\Omega)$, $\tilde{h}_3 \in H^{3/2}(\Omega)$, $d_t \in L_2(S_2)$, $v'd_{x'} \in L_2(S_2)$, $f_3 \in L_2(S_2)$, $F \in L_2(\Omega)$, *and* $\left| \int_{S_2} v_3 h_3 k_3 dS_2 \right| < \infty$. *Then*

$$\|k\|_{1,\Omega}^2 + \sum_{\alpha=1}^{2} |k \cdot \bar{\tau}_\alpha|_{2,S_2}^2 \le c\Bigg(\|\tilde{h}\|_{1,\Omega}^2 + |\tilde{h}_{3,x_3}|_{2,S_2}^2 + |d_t|_{2,S_2}^2$$

$$+ |v'd_{x'}|_{2,S_2}^2 + |f_3|_{2,S_2}^2 + |F|_{2,\Omega}^2 + \bigg| \int_{S_2} v_3 h_3 k_3 dS_2 \bigg| \Bigg),$$

$$(9.31)$$

$$\|h\|_{1,\Omega}^2 + \sum_{\alpha=1}^{2} |h \cdot \bar{\tau}_\alpha|_{2,S_1}^2 \le c\Bigg(\|\tilde{h}\|_{1,\Omega}^2 + \|\tilde{h}_3\|_{3/2,\Omega}^2 + |d_t|_{2,S_2}^2 + |v'd_{x'}|_{2,S_2}^2$$

$$+ |f_3|_{2,S_2}^2 + |F|_{2,\Omega}^2 + \bigg| \int_{S_2} v_3 h_3 k_3 dS_2 \bigg| \Bigg).$$

Proof Multiplying $(9.18)_1$ by k and integrating over Ω yield

$$\int_\Omega \operatorname{div} \mathbb{T}(h, q) \cdot k dx = \int_\Omega F \cdot k dx. \qquad (9.32)$$

Integrating by parts in the term on the l.h.s. we obtain

$$\int_\Omega \operatorname{div} \mathbb{T}(h, q) \cdot k dx = \int_{S_1} \bar{n} \cdot \mathbb{T}(h, q) \cdot \bar{\tau}_\alpha k \cdot \bar{\tau}_\alpha dS_1 + \int_{S_2} \mathbb{T}_{33}(h, q) k_3 dS_2$$

$$- \frac{\nu}{2} \int_\Omega \mathbb{D}(h) \cdot \mathbb{D}(k) dx = -\gamma \int_{S_1} h \cdot \bar{\tau}_\alpha k \cdot \bar{\tau}_\alpha dS_1 + \int_{S_2} \mathbb{T}_{33}(h, q) k_3 dS_2$$

$$- \frac{\nu}{2} |\mathbb{D}(k)|_{2,\Omega}^2 - \frac{\nu}{2} \mathbb{D}(\tilde{h}) \cdot \mathbb{D}(k) dx.$$

Employing the results in (9.32) gives

$$\frac{\nu}{2} |\mathbb{D}(k)|_{2,\Omega}^2 + \gamma |k \cdot \bar{\tau}_\alpha|_{2,S_1}^2 = -\gamma \int_{S_1} \tilde{h} \cdot \bar{\tau}_\alpha k \cdot \bar{\tau}_\alpha dS_1$$

$$- \frac{\nu}{2} \int_\Omega \mathbb{D}(\tilde{h}) \cdot \mathbb{D}(k) dx + \int_{S_2} \mathbb{T}_{33}(h, q) \cdot k_3 dS_2 - \int_\Omega F \cdot k dx.$$

$$(9.33)$$

Applying the Hölder and Young inequalities to the first two terms on the r.h.s. of (9.33) implies

$$\frac{\nu}{2} |\mathbb{D}(k)|_{2,\Omega}^2 + \gamma |k \cdot \bar{\tau}_\alpha|_{2,S_1}^2 \le \frac{\varepsilon_1}{2} |\mathbb{D}(k)|_{2,\Omega}^2 + \frac{1}{2\varepsilon_1} \frac{\nu^2}{4} |\mathbb{D}(\tilde{h})|_{2,\Omega}^2$$

$$+ \frac{\varepsilon_2}{2} |k \cdot \bar{\tau}_\alpha|_{2,S_1}^2 + \frac{1}{2\varepsilon_2} \gamma^2 |\tilde{h} \cdot \bar{\tau}_\alpha|_{2,S_1}^2 + \int_{S_2} \mathbb{T}_{33}(h, q) k_3 dS_2 - \int_\Omega F \cdot k dx.$$

Setting $\varepsilon_1 = \frac{\nu}{2}$ and $\varepsilon_2 = \gamma$ we derive

$$
|\mathbb{D}(k)|_{2,\Omega}^2 + \sum_{\alpha-1}^{2} |k \cdot \bar{\tau}_\alpha|_{2,S_1}^2 \leq c\|\tilde{h}\|_{1,\Omega}^2 + \varepsilon\|k\|_{1,\Omega}^2
$$
$$
+ c(1/\varepsilon)\left(\left|\int_{S_2} \mathbb{T}_{33}k_3 dS_2\right| + |F|_{2,\Omega}^2\right). \tag{9.34}
$$

The third component of the Navier-Stokes equations (1.1) projected on S_2 takes the form

$$
d_t + v' \cdot d_{x'} + v_3 h_3 - \nu(\Delta' d + h_{3,x_3}) + q = f_3.
$$

Hence

$$
\mathbb{T}_{33}(h,q)|_{S_2} = d_t + v' \cdot d_{x'} + v_3 h_3 - f_3.
$$

Using that ε is sufficiently small, the formula for $\mathbb{T}_{33}$ in (9.34) and (9.19) implies the inequality

$$
\|k\|_{1,\Omega}^2 + \sum_{\alpha=1}^{2} |k \cdot \bar{\tau}_\alpha|_{2,S_1}^2 \leq c\|\tilde{h}\|_{1,\Omega}^2 + c(|d_t|_{2,S_2}^2 + |v'd_{x'}|_{2,S_2}^2
$$
$$
+ |f_3|_{2,S_2}^2 + |F|_{2,\Omega}^2) + \left|\int_{S_2} v_3 h_3 k_3 dS_2\right|. \tag{9.35}
$$

In view of relation (9.17) we have

$$
\|h\|_{1,\Omega}^2 + \sum_{\alpha=1}^{2} |h \cdot \bar{\tau}_\alpha|_{2,S_1}^2 \leq c(\|\tilde{h}\|_{1,\Omega}^2 + \|\tilde{h}_3\|_{3/2,\Omega}^2)
$$
$$
+ c(|d_t|_{2,S_2}^2 + |v'd_{x'}|_{2,S_2}^2 + |f_3|_{2,S_2}^2 + |F|_{2,\Omega}^2) + \left|\int_{S_2} v_3 h_3 k_3 dS_2\right|. \tag{9.36}
$$

Hence, (9.35) and (9.36) imply (9.31). This concludes the proof. □

Lemma 9.4 *Assume that* $d_{x'} \in H^1(S_2)$, $d_t \in L_2(S_2)$, $v' \in H^1(\Omega)$, $f_3 \in L_2(S_2)$, $F \in L_2(\Omega)$, $d \in L_2(S_2)$, *and* $h_3 \in L_2(\Omega)$. *Then the following inequality holds*

$$
\|h\|_{2,\Omega} + \|q\|_{1,\Omega} \leq c(\|d_{x'}\|_{1,S_2} + |d_t|_{2,S_2} + \|v'\|_{1,\Omega}^2\|d_{x'}\|_{1,S_2}^2
$$
$$
+ |f_3|_{2,S_2} + |F|_{2,\Omega} + |v_3|_{2,S_2}^2|h_3|_{2,\Omega}^2 + |d|_{2,S_2}^2\|d_{x'}\|_{1,S_2}^2). \tag{9.37}
$$

Proof To prove the lemma we use the partition of unity introduced in Chap. 2. Let $k \in \mathfrak{M}$. Then problem (9.1) restricted to neighborhoods $\Omega^{(k)} = \text{supp}\,\zeta^{(k)}$, $k \in \mathfrak{M}$, has the form

$$\nu \Delta h - \nabla q = F,$$
$$\text{div}\, h = 0. \tag{9.38}$$

Using notation $h^{(k)} = h\zeta^{(k)}$, $q^{(k)} = q\zeta^{(k)}$, and $F^{(k)} = F\zeta^{(k)}$ we transform (9.38) to the following problem for the localized functions

$$\nu \Delta h^{(k)} - \nabla q^{(k)} = F^{(k)} + 2\nu \nabla h \nabla \zeta^{(k)} + \nu h \Delta \zeta^{(k)} - q \nabla \zeta^{(k)},$$
$$\text{div}\, h^{(k)} = h \cdot \nabla \zeta^{(k)}. \tag{9.39}$$

From the theory of stationary Stokes system we have the inequality

$$\|h^{(k)}\|_{2,\Omega} + \|q^{(k)}\|_{1,\Omega} \le c|F^k|_{2,\Omega} + c(\|h\|_{1,\Omega^{(k)}} + |q|_{2,\Omega^{(k)}}). \tag{9.40}$$

Let $k \in \mathfrak{N}_1$ and let $\xi_k \in S_1$. Moreover, $\Omega^{(k)} = \text{supp}\,\zeta^{(k)}$ is a neighborhood of ξ_k such that ξ_k is the middle point of $\bar{\Omega}^{(k)} \cap S_1$. We assume also that $\Omega^{(k)}$ is located in a positive distance from the edges L. Introduce a new coordinate system $y = (y_1, y_2, y_3)$ with origin at $\xi^{(k)}$ and $y = Y_k(x)$, where $x = (x_1, x_2, x_3)$ is the global Cartesian system in Ω and Y_k is a composition of a rotation and a translation. Let us assume that $\Omega^{(k)} \cap S_1$ is described in the form

$$y_1 = F_k(y_1, y_3).$$

Then we introduce new coordinates

$$z_2 = y_2, \quad z_3 = y_3,$$
$$z_1 = y_1 - F(y_2, y_3). \tag{9.41}$$

The transformation (9.41) is denoted by

$$z = \Psi_k(y) = \Psi_k(Y_k(x)) \equiv \Phi_k(x), \quad z = (z_1, z_2, z_3).$$

In these coordinates z we have

$$\text{supp}\,\zeta^{(k)} \cap S_1 \subset \{z \colon z_1 = 0\}.$$

We restrict problem (9.1) to neighborhood $\Omega^{(k)}$, $k \in \mathfrak{N}_1$ and express it in the new coordinates z. For this we need the notation

$$
\begin{aligned}
\hat{u}(z) &= u(\Phi_k^{-1}(z)), & u^{(k)} &= \hat{u}(z)\hat{\zeta}^{(k)}(z), \\
\hat{n}(z) &= \bar{n}(\Phi_k^{-1}(z)), & \hat{\tau}_\alpha &= \bar{\tau}_\alpha(\Phi_k^{-1}(z)), \\
\hat{\nabla}_z &= \left.\frac{\partial z_i}{\partial x}\right|_{x=\Phi_k^{-1}(z)} \nabla_{z_i}, & \nabla_{z_i} &= \partial_{z_i}.
\end{aligned}
\tag{9.42}
$$

In coordinates (9.41), $\bar{n}_z = (1,0,0)$, $\bar{\tau}_{z_1} = (0,1,0)$, and $\bar{\tau}_{zz} = (0,0,1)$. Then (9.1) in $\Omega^{(k)}$, $k \in \mathfrak{N}_1$, and in coordinates z has the form

$$
\begin{aligned}
\nu\nabla_z^2 h^{(k)} - \nabla_z q^{(k)} &= \nu(\nabla_z^2 - \hat{\nabla}_z^2)h^{(k)} - (\nabla_z - \hat{\nabla}_z)q^{(k)} \\
&\quad + F^{(k)} + 2\nu\hat{\nabla}_z\hat{h}\hat{\nabla}_z\hat{\zeta}^{(k)} + \nu\hat{h}\hat{\nabla}_z^2\hat{\zeta}^{(k)} - \hat{q}\hat{\nabla}_z\hat{\zeta}^{(k)} & z_1 &> 0, \\
\operatorname{div}_z h^{(k)} &= (\operatorname{div}_z - \hat{\operatorname{div}}_z)h^{(k)} + \hat{h}\hat{\nabla}_z\hat{\zeta}^{(k)} & z_1 &> 0, \\
h_1^{(k)} &= h_1^{(k)} - h^{(k)} \cdot \hat{n}(z) & z_1 &= 0, \\
h_{\alpha,z_1}^{(k)} &= h_{\alpha,z_1}^{(k)} - \hat{n}(z) \cdot \hat{\mathbb{D}}_z(h^{(k)}) \cdot \hat{\tau}_\alpha(z) + n_1\widehat{\partial_{x_i}\zeta^{(k)}}h_j\tau_{j\alpha} \\
&\quad + n_i\widehat{\partial_{x_j}\zeta^{(k)}}h_i\tau_{j\alpha} - \hat{\tau}_\alpha \cdot \hat{\nabla}_z h_1^{(k)} & z_1 &= 0.
\end{aligned}
\tag{9.43}
$$

Hence, (9.43) is the stationary Stokes system in the half space $z_1 > 0$ with the Dirichlet and Neumann conditions on $z_1 = 0$ for coordinates of $h^{(k)}$. From the theory of the Stokes system and sufficiently small λ, we get

$$
\begin{aligned}
\|h^{(k)}\|_{2,\hat{\Omega}^{(k)}} + \|q^{(k)}\|_{1,\hat{\Omega}^{(k)}} &\le c(|F^{(k)}|_{2,\hat{\Omega}^{(k)}} + \|\hat{h}\|_{1,\hat{\Omega}^{(k)}} \\
&\quad + |\hat{q}|_{2,\hat{\Omega}^{(k)}}), \quad k \in \mathfrak{N}_1,
\end{aligned}
\tag{9.44}
$$

where $\hat{\Omega}^{(k)} = \Phi_k\Omega^{(k)}$.

Let $k \in \mathfrak{N}_2$ and $\xi^{(k)} \in S_2$. We assume that $\xi^{(k)}$ is the middle point of $\operatorname{supp}\zeta^{(k)} \cap S_2$. Since S_2 is flat we introduce the following new coordinates z such that $z_3 = x_3$, $z_i = x_i + a_i$, and $i = 1, 2$, where a_i are constants and describe a translation. In this case we localize problem (9.1) to the following one

$$
\begin{aligned}
\nu\nabla_z^2 h^{(k)} - \nabla_z q^{(k)} &= F^{(k)} + 2\nu\nabla_z\hat{\zeta}^{(k)}\nabla_z\hat{h} + \nu\hat{h}\hat{\nabla}_z\hat{\zeta}^{(k)} & z_3 &> 0, \\
\operatorname{div}_z h^{(k)} &= \hat{h} \cdot \nabla_z\hat{\zeta}^{(k)} & z_3 &> 0, \\
h_i^{(k)} &= -\hat{d}_{,z_i}\hat{\zeta}^{(k)} & z_3 &= 0, \\
h_{3,z_3}^{(k)} &= \hat{h}_3\hat{\zeta}_{,z_3}^{(k)} + \Delta_z'\hat{d}\hat{\zeta}^{(k)} & z_3 &= 0.
\end{aligned}
\tag{9.45}
$$

For solutions to (9.45) the following estimate holds

$$\|h^{(k)}\|_{2,\hat{\Omega}^{(k)}} + \|q^{(k)}\|_{1,\hat{\Omega}^{(k)}} \le c(\|\hat{h}\|_{1,\hat{\Omega}^{(k)}} + |F^{(k)}|_{2,\hat{\Omega}^{(k)}}$$
$$+ \|\hat{d}_{,z}\|_{3/2,\hat{S}_2^{(k)}}). \tag{9.46}$$

Finally, we pass to the most difficult case—neighborhoods of edges. Let $k \in \mathfrak{N}_3$, $\xi^{(k)} \in L$, and $\Omega^{(k)} = \text{supp}\,\zeta^{(k)}$ is a neighborhood of $\xi^{(k)}$. We introduce new coordinates $y = (y_1, y_2, y_3)$ such that $\Omega^{(k)} \cap S_1 = \{y: y_1 = F_k(y_2, y_3)\}$, $\Omega^{(k)} \cap S_2 \subset \{y: y_3 = 0\}$, and $L = \{y: y_1 = F_k(y_2, 0)\}$.

Next we define coordinate z by the relations $z_3 = y_3$, $z_1 = y_1 - F_k(y_2, y_3)$, and $z_2 = y_2$.

Then the localized problem (9.1) in the right dihedral angle has the form

$$\nu\nabla_z^2 h^{(k)} - \nabla_z q^{(k)} = \nu(\nabla_z^2 - \hat{\nabla}_z^2)h^{(k)} - (\nabla_z - \hat{\nabla}_z)q^{(k)}$$

$$+F^{(k)} + 2\nu\hat{\nabla}_z\hat{h}\hat{\nabla}_z\hat{\zeta}^{(k)} + \nu\hat{h}\hat{\nabla}_z^2\hat{\zeta}^{(k)} - \hat{q}\hat{\nabla}_z\hat{\zeta}^{(k)} \quad z_1 > 0, \quad z_3 > 0,$$

$$\text{div}\,_z h^{(k)} = (\text{div}\,_z - \hat{\text{div}}\,_z)h^{(k)} + \hat{h}\hat{\nabla}_z\hat{\zeta}^{(k)} \quad z_1 > 0, \quad z_3 > 0,$$

$$h_1^{(k)} = h_1^{(k)} - h^{(k)} \cdot \hat{n}(z) \equiv G_1^{(k)} \quad z_1 = 0$$

$$h_{\alpha,z_1}^{(k)} = -h_{1,z_\alpha}^{(k)} + \bar{n}_z \cdot \mathbb{D}(h^{(k)}) \cdot \bar{\tau}_{z_\alpha} - \hat{n}(z)\hat{\mathbb{D}}_z(h^{(k)}) \cdot \hat{\tau}_\alpha(z) \tag{9.47}$$

$$+ n_i\widehat{\partial_{x_i}\zeta^{(k)}}h_j\tau_{j\alpha} + n_i\widehat{\partial_{x_j}\zeta^{(k)}}h_i\tau_{j\alpha} \equiv G_\alpha^{(k)}, \quad \alpha = 1, 2, \quad z_1 = 0,$$

$$h_i^{(k)} = -\hat{d}_{z_i}\hat{\zeta}^{(k)} \quad i = 1, 2, \quad z_3 = 0,$$

$$h_{3,z_3}^{(k)} = \hat{h}_3\hat{\zeta}_{,z_3}^{(k)} + \Delta_z'\hat{d}\hat{\zeta}^{(k)} \quad z_3 = 0,$$

where $\bar{n}_z = (1, 0, 0)$, $\bar{\tau}_{z1} = (0, 1, 0)$, and $\bar{\tau}_{z2} = (0, 0, 1)$ on $\{z: z_1 = 0\}$.

We extend the functions $G_j^{(k)}$, $j = 1, 2, 3$, on $z_1 > 0$ and denote the extended functions by $\bar{G}_j^{(k)}$, $j = 1, 2, 3$. Introducing the new functions $g^{(k)} = h^{(k)} - \bar{G}^{(k)}$ we obtain the problem (9.47) with vanishing boundary conditions on $z_1 = 0$. Next after reflection with respect to the plane $z_1 = 0$ we obtain problem (9.47) in the half space $z_3 > 0$. Then for λ sufficiently small, we obtain the inequality for solutions to problem (9.47)

$$\|h^{(k)}\|_{2,\Omega} + \|q^{(k)}\|_{1,\Omega} \le c(\|\hat{h}\|_{1,\hat{\Omega}^{(k)}} + |\hat{F}^{(k)}|_{2,\hat{\Omega}^{(k)}}$$
$$+ |\hat{q}|_{2,\hat{\Omega}^{(k)}}), \quad k \in \mathfrak{N}_3. \tag{9.48}$$

We pass to variables x in (9.44), (9.46), and (9.48). Next adding the transformed inequalities (9.44), (9.46) and (9.48) to (9.40), summing up over all neighborhoods of the partition of unity, we derive the inequality

$$\|h\|_{2,\Omega} + \|q\|_{1,\Omega} \le c(\|h\|_{1,\Omega} + |q|_{2,\Omega} + |F|_{2,\Omega}). \tag{9.49}$$

To estimate the second norm on the r.h.s. we consider the system

$$\begin{aligned} \nu \Delta h - \nabla q &= F, \\ \operatorname{div} h &= 0. \end{aligned} \tag{9.50}$$

Let φ be a function such that

$$\operatorname{div} \varphi = q, \quad \varphi|_S = 0, \tag{9.51}$$

and let $q \in L_2(\Omega)$. Then the paper by Kapitanskii and Pileckas [KP] implies the existence of φ such that $\varphi \in H^1(\Omega)$ and

$$\|\varphi\|_{1,\Omega} \le c|q|_{2,\Omega}. \tag{9.52}$$

Multiplying (9.50) by φ, integrating over Ω and by parts, and using (9.51), (9.52) we have

$$|q|_{2,\Omega} \le c(|F|_{2,\Omega} + \|h\|_{1,\Omega}). \tag{9.53}$$

Employing (9.53) in (9.49) and next exploiting (9.31)$_2$ we obtain

$$\begin{aligned} \|h\|_{2,\Omega} + \|q\|_{1,\Omega} \le c\bigg(&\|\tilde{h}\|_{1,\Omega} + \|\tilde{h}_3\|_{3/2,\Omega} + |d_t|^2_{2,S_2} + |v'd_{x'}|^2_{2,S_2} \\ &+ |f_3|^2_{2,S_2} + |F|^2_{2,\Omega} + \bigg| \int_{S_2} v_3 h_3 k_3 dS_2 \bigg| \bigg). \end{aligned} \tag{9.54}$$

The last term on the r.h.s. of (9.54) is estimated by

$$\bigg| \int_{S_2} v_3 h_3^2 dS_2 \bigg| + \bigg| \int_{S_2} v_3 h_3 \tilde{h}_3 dS_2 \bigg| \le |d|_{2,S_2} |h_3|^2_{4,S_2} + |d|_{2,S_2} |h_3|_{4,S_2} |\tilde{h}_3|_{4,S_2}$$

$$\le \varepsilon \|h_3\|^2_{2,\Omega} + c(1/\varepsilon)|v_3|^2_{2,S_2} |h_3|^2_{2,\Omega} + \varepsilon |h_3|^2_{4,S_2} + c(1/\varepsilon)|d|_{2,S_2} \|\tilde{h}_3\|_{1,\Omega}.$$

Moreover, the fourth term on the r.h.s. of (9.54) is estimated by

$$|v'|^2_{4,S_2} |d_{x'}|^2_{4,S_2} \le \|v'\|^2_{1,\Omega} \|d_{x'}\|^2_{1,S_2}.$$

Employing the estimates in (9.54) and assuming that ε is sufficiently small yield (9.37) and concludes the proof.   □

Chapter 10
The Neumann Problem (3.6) in L_2-Weighted Spaces

Abstract In this chapter and Chap. 11 we derive weighted estimates for solutions to the Neumann problem for the Poisson equation (3.6), i.e.

$$\Delta\varphi = -\operatorname{div} b \quad \text{in} \quad \Omega,$$

$$\bar{n} \cdot \nabla\varphi = 0 \quad \text{on} \quad S,$$

$$\int_\Omega \varphi dx = 0.$$

These estimates are necessary to prove global energy estimates formulated in Lemma 3.2. In this chapter we apply L_2 approach to get H^2 weighted regularity and for this we need weighted energy type estimates for the Fourier transform in the directions perpendicular to the x_3-axis. The results are formulated in Lemmas 10.4–10.6. We get the estimate in a neighborhood of S_2 and use a partition of unity to obtain an estimate for φ in whole Ω.

In this chapter and Chap. 11 we analyze the problem (3.6) to achieve a weighted estimate for solutions in a neighborhood of S_2 which makes possible to prove the energy estimate established in Lemma 3.2. However, the techniques and results of these chapters are totally different. In Chap. 10 we derive the weighted estimate in the L_2-spaces using the Fourier transform in the directions perpendicular to the x_3-axis and some delicate weighted estimates possible in the L_2-approach only (see Kubica and Zajączkowski [KZ], [Z9]). However, to prove Lemma 3.2 we need L_p estimate with $p \geq 3$. Therefore, in Chap. 11 we show L_p-weighted estimate using the estimate proved in Chap. 10. We apply the local regularity technique connected with the considered weights (see Maz'ya and Plamenevskii [MP]). Hence we cannot mix the above different approaches. Estimated in weighted norms can also be found in [CF].

© Springer Nature Switzerland AG 2019

J. Rencławowicz, W. M. Zajączkowski, *The Large Flux Problem to the Navier-Stokes Equations*, Advances in Mathematical Fluid Mechanics, https://doi.org/10.1007/978-3-030-32330-1_10

In this chapter we examine problem (3.6) in the form

$$- \Delta\varphi = f \qquad \text{in } \Omega,$$
$$\bar{n} \cdot \nabla\varphi = 0 \qquad \text{on } S,$$
$$\int_\Omega \varphi \, dx = 0,$$

(10.1)

where

$$f = \text{div } b$$

(10.2)

and b is defined by (3.4).

We assume that the considered domain has the form as in Chap. 1 (Fig. 10.1).

For convenience we replace $-a$ with 0 in this and the following chapter. The shape of Ω is appropriate for application of weighted spaces. Since the weighted spaces have the weight as a power function of the distance from either $S_2(0)$ or $S_2(a)$ it is convenient to examine problem (10.1) in a neighborhood of $S_2(0)$ only, because considerations near $S_2(a)$ are similar. To consider problem (10.1) in a neighborhood of $S_2(0)$ we introduce a smooth cut-off function $\zeta = \zeta(x_3)$ such that $\zeta(x_3) = 1$ for $x_3 \leq r$ and $\zeta(x_3) = 0$ for $x_3 \geq \varrho$, $r < \varrho < a$. Let

$$\tilde{\varphi} = \varphi\zeta, \quad \tilde{f} = f\zeta.$$

(10.3)

Then $\tilde{\varphi}$ is a solution to the problem

$$- \Delta\tilde{\varphi} = \tilde{f} - 2\nabla\varphi\nabla\zeta - \varphi\Delta\zeta \equiv \tilde{f}'$$
$$\bar{n} \cdot \nabla\tilde{\varphi}|_{S_2(0)} = 0,$$
$$\bar{n} \cdot \nabla\tilde{\varphi}|_{S_1} = 0,$$
$$\tilde{\varphi}|_{x_3=\varrho} = 0,$$

(10.4)

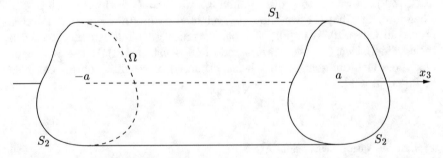

Fig. 10.1 Cylindrical domain Ω

where we used that $\bar{n}|_{S_1}$ does not have the third coordinate. Let

$$\varphi(0) = \varphi|_{x_3=0} \quad \text{and} \quad \tilde{\varphi}(0) = \varphi(0)\zeta \tag{10.5}$$

and

$$\varphi' = \tilde{\varphi} - \tilde{\varphi}(0). \tag{10.6}$$

Moreover, $\bar{n} \cdot \nabla \tilde{\varphi}(0)|_{S_2(0)} = 0$, $\bar{n} \cdot \nabla \varphi(0)|_{S_1} = 0$ because $\bar{n} \cdot \nabla \varphi|_{S_1} = 0$. Since $\bar{n} \cdot \nabla \zeta|_{S_1} = 0$ we have that $\bar{n} \cdot \nabla \tilde{\varphi}(0)|_{S_1} = 0$. Therefore φ' is a solution to the problem

$$\begin{aligned}
-\Delta\varphi' &= \tilde{f}' + \Delta\tilde{\varphi}(0) \equiv f', \\
\bar{n} \cdot \nabla\varphi'|_{S_2(0)} &= 0, \\
\bar{n} \cdot \nabla\varphi'|_{S_1} &= 0, \\
\varphi'|_{x_3=\varrho} &= 0.
\end{aligned} \tag{10.7}$$

To examine the problem (10.7) in weighted spaces we introduce the following.

Definition 10.1 Let $\Omega_\varrho = \{x \in \Omega : 0 < x_3 < \varrho\}$ and

$$H^k_\mu(\Omega_\varrho) = \left\{ u : \|u\|_{H^k_\mu(\Omega_\varrho)} = \left(\sum_{|\alpha| \le k} \int_{\Omega_\varrho} dx' dx_3 |D^\alpha_x u|^2 x_3^{2(\mu+|\alpha|-k)} \right)^{1/2} < \infty \right\},$$

where $k \in \mathbb{N}_0 \equiv \mathbb{N} \cup \{0\}$, $\mu \in \mathbb{R}$, $\alpha = (\alpha_1, \alpha_2, \alpha_3)$ is a multiindex, $|\alpha| = \alpha_1 + \alpha_2 + \alpha_3$, $\alpha_i \in \mathbb{N}_0$, $i = 1, 2, 3$, $D^\alpha_x = \partial^{\alpha_1}_{x_1} \partial^{\alpha_2}_{x_2} \partial^{\alpha_3}_{x_3}$. Moreover, we denote $L_{2,\mu}(\Omega_\varrho) = H^0_\mu(\Omega_\varrho)$.

Lemma 10.2 *Assume that $f' \in L_{2,\mu}(\Omega_\varrho)$, $\mu \in (0,1)$, $\varphi(0) \in L_2(\Omega)$, and ζ is the cut-off function introduced above (10.3). Then solutions to (10.7) satisfy*

$$\|\varphi'\|_{H^1(\Omega)} + \|\varphi'\|_{L_{2,-\mu}(\Omega)} \le c(\|f'\|_{L_{2,\mu}(\Omega)} + \|\varphi(0)\dot{\zeta}\|_{L_2(\Omega)}). \tag{10.8}$$

Proof Multiplying (10.7)$_1$ by φ', integrating over Ω and by parts, and using boundary conditions yield

$$\int_\Omega |\nabla\varphi'|^2 dx = \int_\Omega f' \cdot \varphi' dx \le \left(\int_\Omega |f'|^2 x_3^{2\mu} dx \right)^{1/2} \left(\int_\Omega |\varphi'|^2 x_3^{-2\mu} dx \right)^{1/2}, \tag{10.9}$$

and in reality all integrals are over Ω_ϱ because functions in (10.9) vanish for $x_3 > \varrho$. We calculate

$$\tilde{\varphi}(x', x_3) - \tilde{\varphi}(x', s) = \int_s^{x_3} \tilde{\varphi}_{,s'}(x', s')ds',$$

where $x' = (x_1, x_2)$. By the Hölder inequality

$$\frac{|\tilde{\varphi}(x', x_3) - \tilde{\varphi}(x', s)|^2}{|x_3 - s|} \leq \int_s^{x_3} |\tilde{\varphi}_{,s'}(x', s')|^2 ds'.$$

Setting $s = 0$ and integrating the inequality over the following set: $\Omega' = \{x \in \Omega: x_3 = \text{const} \in [0, a]\}$, we get

$$\int_{\Omega'} \frac{|\tilde{\varphi}(x', x_3) - \tilde{\varphi}(x', 0)|^2}{x_3} dx' \leq \int_{\Omega'} dx' \int_0^{x_3} |\tilde{\varphi}_{,s}(x', s)|^2 ds, \qquad (10.10)$$

where the l.h.s. of (10.10) equals

$$\int_{\Omega'} \frac{|\varphi'(x', x_3)|^2}{x_3} dx'.$$

Now, we consider the second factor on the r.h.s. of (10.9). We have for $\mu \in [0, 1)$ the inequalities

$$\int_{\Omega} |\varphi'|^2 x_3^{-2\mu} dx = \int_{\Omega'} dx' \int_0^a \frac{|\varphi'|^2}{x_3} x_3^{1-2\mu} dx_3$$

$$\leq \int_{\Omega'} dx' \sup_{x_3 \leq a} \frac{|\varphi'|^2}{x_3} \int_0^a x_3^{1-2\mu} dx_3$$

$$\leq \int_0^a x_3^{1-2\mu} dx_3 \int_{\Omega'} dx' \int_0^a |\tilde{\varphi}_{,x_3'}(x', x_3)|^2 dx_3' \qquad (10.11)$$

$$\leq c \int_{\Omega'} dx' \int_0^a |\tilde{\varphi}_{,x_3'}(x', x_3')|^2 dx_3' \leq c \int_{\Omega'} dx' \int_0^a |\varphi'_{,x_3}(x', x_3)|^2 dx_3$$

$$+ c \int_{\Omega'} dx' \int_0^a |\tilde{\varphi}(0)_{,x_3}|^2 dx_3,$$

where the last term on the r.h.s. of the above inequality equals

$$c \int_{\Omega'} dx' \int_{\text{supp}\,\zeta} |\varphi(x', 0)|^2 |\dot{\zeta}|^2 dx_3,$$

where $\dot{\zeta} = \zeta_{,x_3}$ and $\text{supp}\,\dot{\zeta} \subset [r, \varrho]$. Employing (10.11) in (10.9) yields

$$\int_{\Omega} |\nabla \varphi'|^2 dx \leq c \int_{\Omega} |f'|^2 x_3^{2\mu} dx + c \int_{\Omega'} dx' \int_{\text{supp}\,\zeta} |\varphi(x', 0)|^2 |\dot{\zeta}|^2 dx_3. \qquad (10.12)$$

In view of (10.9)–(10.12) and the Poincaré inequality we derive (10.8). This concludes the proof. □

Remark 10.3 Estimate (10.8) describes local properties of solutions to problem (10.1) in a neighborhood of $S_2(0)$. We derive a similar estimate for φ near $S_2(a)$.

To obtain an estimate for φ in whole Ω we need a partition of unity of Ω given as $\{\zeta^{(k)}(x_3)\}_{k=1,\ldots,N_0}$, $\sum_{k=1}^{N_0} \zeta^{(k)}(x_3) = 1$, where $\zeta^{(1)}(x_3)$ is equal to $\zeta(x_3)$ introduced above. Similarly $\zeta^{N_0}(x_3)$ is equal to 1 near $S_2(a)$. Hence, we can derive (10.8) in any supp $\zeta^{(k)}$, $k = 2, \ldots, N_0 - 1$, also, where the weighted spaces are not needed.

Adding the estimates we obtain

$$
\sum_{k=1,N_0} \left(\|\varphi'\|_{H^1(\Omega^{(k)})} + \|\varphi'\|_{L_{2,-\mu}(\Omega^{(k)})} \right) + \sum_{k=2}^{N_0-1} \|\varphi\|_{H^1(\Omega^{(k)})}
$$
$$
\leq c \sum_{k=1,N_0} \|f'\|_{L_{2,\mu}(\Omega^{(k)})} + c \sum_{k=2}^{N_0-1} \|f\|_{L_2(\Omega^{(k)})} \tag{10.13}
$$
$$
+ c \sum_{k=1,N_0} \|\varphi(x',0)\dot{\zeta}^{(k)}\|_{L_2(\Omega)},
$$

where $\Omega^{(k)} = \text{supp}\,\zeta^{(k)}$ and in supp $\zeta^{(N_0)}$ there are introduced such local coordinates that $S_2(a)$ is determined by $x_3 = 0$.

Having the estimate (10.13), the existence of solutions to (10.1) follows from the first Fredholm theorem.

Applying local considerations, we increase the regularity of weak solutions described by Remark 10.3. The most difficult considerations are expected in neighborhoods $\Omega^{(1)}$ and $\Omega^{(N_0)}$ because in these domains the technique of weighted spaces must be used. We restrict our considerations to neighborhoods of the edge $L(0)$. Let $\xi_1, \ldots, \xi_{M_0}$ belong to $L(0)$. Take smooth cut-off functions $\eta^{(1)}, \ldots, \eta^{(M_0)}$ such that ξ_k is the middle point of $L(0) \cap \overline{\text{supp}\,\eta^{(k)}}$. Assume that $\{\eta^{(k)}\}_{k=1,\ldots,M_0}$ is a partition of unity in a neighborhood of $L(0)$. Hence $\bigcup_{k=1}^{M_0} \text{supp}\,\eta^{(k)}$ cover whole $L(0)$. Consider (10.7) in supp $\eta^{(k)} \subset$ supp $\zeta^{(1)}$. Introduce new local coordinates with origin at ξ_k. Denote them by $y = (y_1, y_2, y_3)$. They are obtained from global coordinates $x = (x_1, x_2, x_3)$ by a translation and a rotation. We denote the transformation by: $y = Y_k(x)$. Next we introduce new coordinates $z = (z_1, z_2, z_3)$ transforming neighborhood supp $\eta^{(k)}$ of point ξ_k into a right dihedral angle. We make such transformation that $S_2(0) \cap \text{supp}\,\eta^{(k)}$ becomes a subset of $\{z : z_3 = 0\}$, $S_1 \cap \overline{\text{supp}\,\eta^{(k)}}$ becomes a subset of $\{z : z_1 = 0\}$, and $L(0) \cap \text{supp}\,\eta^{(k)}$ is transformed into z_2-line which is the intersection of planes $z_1 = 0$ and $z_3 = 0$.

We denote the transformation by

$$z = \Phi_k(y) = \Phi_k(Y_k(x)) \equiv \Psi_k(x).$$

Since the index is fixed we omit it in the next considerations.

Introducing the notation

$$\varphi''(z) = \varphi'(\Psi^{-1}(z)), \quad f''(z) = f'(\Psi^{-1}(z)) \tag{10.14}$$

we express problem (10.7) in the form

$$
\begin{aligned}
-\nabla_z^2 \varphi'' &= -\nabla_z^2 \varphi'' + \nabla_\Psi^2 \varphi'' + f'' \equiv \tilde{f}, \quad z_3 > 0, \quad z_1 > 0, \\
\varphi''_{,z_3} &= 0 \quad z_3 = 0, \tag{10.15} \\
\varphi''_{,z_1} &= 0 \quad z_1 = 0,
\end{aligned}
$$

where $\nabla_\Psi = \frac{\partial z}{\partial x}\big|_{x=\Psi^{-1}(z)} \nabla_z$. We extend φ'' by 0 for $z_3 > \Psi(\varrho)$. Next we make the reflection with respect to the plane $z_1 = 0$ such that the reflected function $\tilde{\varphi}''$ satisfies

$$
\begin{aligned}
\tilde{\varphi}''(z_1, z_2, z_3) &= \varphi''(z_1, z_2, z_3), \quad z_1 > 0, \\
\tilde{\varphi}''(z_1, z_2, z_3) &= \varphi''(-z_1, z_2, z_3) \quad z_1 < 0. \tag{10.16}
\end{aligned}
$$

Hence, after an extension by zero, the reflected solution to (10.15) satisfies

$$
\begin{aligned}
-\nabla_z^2 \tilde{\varphi}'' &= \tilde{f} \quad z_3 > 0, \\
\tilde{\varphi}''_{,z_3} &= 0 \quad z_3 = 0, \tag{10.17}
\end{aligned}
$$

where $\tilde{f}$ has a compact support.

For further considerations it is convenient to write (10.17) in the form

$$
\begin{aligned}
-\Delta_x u &= f \quad x_3 > 0, \\
u_{,x_3} &= 0 \quad x_3 = 0, \tag{10.18} \\
u &= 0 \quad x_3 = a.
\end{aligned}
$$

Let us consider the Fourier transform

$$\tilde{u}(\xi, x_3) = \int_{\mathbb{R}^2} e^{-i\xi \cdot x'} u(x', x_3) dx', \tag{10.19}$$

where $\xi = (\xi_1, \xi_2)$, $x \cdot \xi = x_1\xi_1 + x_2\xi_2$. Applying this transformation to (10.18) yields

$$-\frac{d^2\tilde{u}}{dx_3^2} + \xi^2\tilde{u} = \tilde{f} \quad x_3 > 0,$$

$$\tilde{u}_{,x_3} = 0 \quad x_3 = 0, \tag{10.20}$$

$$\tilde{u} = 0 \quad x_3 = a.$$

We set

$$\tilde{u}|_{x_3=0} = \tilde{u}(0), \quad \hat{u} = \tilde{u} - \tilde{u}(0)\zeta(x_3), \quad \hat{u}_{,x_3}|_{x_3=0} = 0, \quad \hat{u}|_{x_3=a} = 0,$$

where ζ is introduced above and $\operatorname{supp}\zeta \subset (0,\varrho)$, $\varrho < a$.

Then $\hat{u}$ is a solution to the problem

$$-\hat{u}_{,x_3x_3} + \xi^2\hat{u} = \tilde{f} + \xi^2\tilde{u}(0)\zeta(x_3) - \tilde{u}(0)\ddot{\zeta}(x_3) \equiv \hat{f}, \quad x_3 > 0,$$

$$\hat{u}_{,x_3} = 0, \quad x_3 = 0, \tag{10.21}$$

$$\hat{u} = 0, \quad x_3 = a.$$

Lemma 10.4 *Assume that*

$$\int_{\mathbb{R}^2} d\xi \int_0^a |\hat{f}|^2 x_3^{2\mu}dx_3 < \infty,$$

$$\int_{\mathbb{R}^2} \xi^2 d\xi \int_0^a |\hat{u}|^2 x_3^{2\mu-2}dx_3 < \infty. \tag{10.22}$$

Then solution $\hat{u}$ to problem (10.21) satisfies

$$\int_{\mathbb{R}^2} \xi^2 d\xi \int_0^a (|\hat{u}_{,x_3}|^2 + \xi^2|\hat{u}|^2)|x_3^{2\mu}dx_3$$

$$+ \int_{\mathbb{R}^2} d\xi \int_0^a (|\partial_{x_3}^2\hat{u}|^2 x_3^{2\mu} + |\partial_{x_3}\hat{u}|^2 x_3^{2\mu-2} + |\hat{u}|^2 x_3^{2\mu-4})dx_3 \tag{10.23}$$

$$\leq c_1 \int_{\mathbb{R}^2} \xi^2 d\xi \int_0^a |\hat{u}|^2 x_3^{2\mu-2}dx_3 + c \int_{\mathbb{R}^2} d\xi \int_0^a |\hat{f}|^2 x_3^{2\mu}dx_3.$$

Proof Multiplying $(10.21)_1$ by $\bar{\hat{u}}x_3^{2\mu}$ and integrating with respect to x_3 over $(0,a)$ we get

$$\int_0^a (-\hat{u}_{,x_3x_3}\bar{\hat{u}}x_3^{2\mu} + \xi^2|\hat{u}|^2)x_3^{2\mu}dx_3 = \int_0^a \hat{f}\bar{\hat{u}}x_3^{2\mu}dx_3,$$

where $\bar{v}$ is the complex conjugate to v. Integrating by parts we obtain

$$\int_0^a (|\hat{u}_{,x_3}|^2 x_3^{2\mu} + \xi^2 |\hat{u}|^2 x_3^{2\mu}) dx_3 = -2\mu \int_0^a \hat{u}_{,x_3} \bar{\hat{u}} x_3^{2\mu-1} dx_3 + \int_0^a \hat{f} \bar{\hat{u}} x_3^{2\mu} dx_3.$$

$$(10.24)$$

By the Hölder and Young inequalities we estimate the first term on the r.h.s. of (10.24) by

$$\frac{\varepsilon_1}{2} \int_0^a |\hat{u}_{,x_3}|^2 x_3^{2\mu} dx_3 + \frac{4\mu^2}{2\varepsilon_1} \int_0^a |\hat{u}|^2 x_3^{2\mu-2} dx_3,$$

and the second by

$$\frac{\varepsilon_2}{2} \xi^2 \int_0^a |\hat{u}|^2 x_3^{2\mu} dx_3 + \frac{1}{2\varepsilon_2} \frac{1}{\xi^2} \int_0^a |\hat{f}|^2 x_3^{2\mu} dx_3.$$

Setting $\varepsilon_1 = \varepsilon_2 = 1$, we obtain from (10.24) the inequality

$$\frac{1}{2} \int_0^a (|\hat{u}_{,x_3}|^2 + \xi^2 |\hat{u}|^2) x_3^{2\mu} dx_3 \leq 2\mu^2 \int_0^a |\hat{u}|^2 x_3^{2\mu-2} dx_3 + \frac{1}{\xi^2} \int_0^a |\hat{f}|^2 x_3^{2\mu} dx_3.$$

We multiply this by $2\xi^2$ and integrate with respect to ξ to get

$$\int_{\mathbb{R}^2} \xi^2 d\xi \int_0^a (|\hat{u}_{,x_3}|^2 + \xi^2 |\hat{u}|^2) x_3^{2\mu} dx_3 \leq 4\mu^2 \int_{\mathbb{R}^2} \xi^2 d\xi \int_0^a |\hat{u}|^2 x_3^{2\mu-2} dx_3$$

$$(10.25)$$

$$+ 2 \int_{\mathbb{R}^2} d\xi \int_0^a |\hat{f}|^2 x_3^{2\mu} dx_3.$$

From (10.21) the following bound follows

$$\|\partial_{x_3}^2 \hat{u}\|_{L_{2,\mu}(0,a)}^2 \leq \xi^4 \|\hat{u}\|_{L_{2,\mu}(0,a)}^2 + \|\hat{f}\|_{L_{2,\mu}(0,a)}^2.$$

Consequently, integrating with respect to ξ implies

$$\int_{\mathbb{R}^2} d\xi \int_0^a |\partial_{x_3}^2 \hat{u}|^2 x_3^{2\mu} dx_3$$

$$(10.26)$$

$$\leq \int_{\mathbb{R}^2} \xi^4 d\xi \int_0^a |\hat{u}|^2 x_3^{2\mu} dx_3 + \int_{\mathbb{R}^2} d\xi \int_0^a |\hat{f}|^2 x_3^{2\mu} dx_3.$$

On the other hand, by the Hardy inequality (see Kondratiev [K, Sect. 4]) we have

$$\int_{\mathbb{R}^2} d\xi \int_0^a |\partial_{x_3}^2 \hat{u}|^2 x_3^{2\mu} dx_3 \geq c \int_{\mathbb{R}^2} d\xi \int_0^a |\partial_{x_3} \hat{u}|^2 x_3^{2(\mu-1)} dx_3$$

$$(10.27)$$

$$\geq c \int_{\mathbb{R}^2} d\xi \int_0^a |\hat{u}|^2 x_3^{2(\mu-2)} dx_3,$$

where we used that $\partial_{x_3} \hat{u}|_{x_3=0} = 0$, $\hat{u}|_{x_3=0} = 0$.

Then inequalities (10.25)–(10.27) imply (10.23). This concludes the proof.
□

We need to estimate the first term on the r.h.s. of (10.23). To this end, we introduce the sets

$$Q_1 = \{(\xi, x_3) \in \mathbb{R}^2 \times (0, a) \colon |\xi|^{-1} x_3^{-1} \le a_1\},$$
$$Q_2 = \{(\xi, x_3) \in \mathbb{R}^2 \times (0, a) \colon |\xi|^{-1} x_3^{-1} \ge a_2\},$$
$$Q_3 = \{(\xi, x_3) \in \mathbb{R}^2 \times (0, a) \colon a_1 \le |\xi|^{-1} x_3^{-1} \le a_2\}.$$

Note that $a_1 < a_2$ are arbitrary, but will be chosen later.

Lemma 10.5 *Assume that*

$$\int_{\mathbb{R}^2} \int_0^a \xi^4 |\hat{u}|^2 x_3^{2\mu} d\xi dx_3 < \infty,$$

$$\int_{\mathbb{R}^2} \int_0^a |\hat{u}|^2 x_3^{2\mu-4} d\xi dx_3 < \infty.$$

Then

$$\int_{\mathbb{R}^2} \xi^2 d\xi \int_0^a |\hat{u}|^2 x_3^{2\mu-2} dx_3 \le a_1^2 \int_{\mathbb{R}^2} \int_0^a \xi^4 |\hat{u}|^2 x_3^{2\mu} d\xi dx_3$$
$$+ \frac{1}{a_2^2} \int_{\mathbb{R}^2} \int_0^a |\hat{u}|^2 x_3^{2\mu-4} d\xi dx_3 + c a_2^{2-2\mu} \int_{\mathbb{R}^2} \int_0^a |\hat{f}|^2 x_3^{2\mu} d\xi dx_3. \tag{10.28}$$

Proof Let us consider the expression

$$\int_{\mathbb{R}^2} \xi^2 d\xi \int_0^a |\hat{u}|^2 x_3^{2\mu-2} dx_3 = \sum_{i=1}^3 \int_{Q_i} \xi^2 |\hat{u}|^2 x_3^{2\mu-2} d\xi dx_3 \equiv \sum_{i=1}^3 I_i,$$

where

$$I_1 \le a_1^2 \int_{Q_1} \xi^4 |\hat{u}|^2 x_3^{2\mu} d\xi dx_3,$$

$$I_2 \le \frac{1}{a_2^2} \int_{Q_2} |\hat{u}|^2 x_3^{2\mu-4} d\xi dx_3,$$

$$I_3 \le a_2^{2-\mu} \int_{Q_3} |\xi|^{4-2\mu} |\hat{u}|^2 d\xi dx_3 \equiv I.$$

To estimate I we recall considerations from [SZ, Sect. 4] and [Z4, (6.14)]. Multiplying (10.21) by $\bar{\hat{u}}$ and integrating over $(0, a)$ we get

$$\int_0^a (|\hat{u}_{,x_3}|^2 + \xi^2|\hat{u}|^2)dx_3$$
$$= \int_0^a \hat{f} \cdot \bar{\hat{u}}dx_3 \leq \left(\int_0^a |\hat{f}|^2 x_3^{2\mu}dx_3 \right)^{1/2} \left(\int_0^a |\hat{u}|^2 x_3^{-2\mu}dx_3 \right)^{1/2}. \tag{10.29}$$

Next, we recall the proof of the interpolation inequality (see [SZ, (2.3)], [Z4, Th. 2.5])

$$\int_{\mathbb{R}_+} |v|^2 x_3^{-2\mu}dx_3 \leq \varepsilon^{2(1-\mu)} \int_{\mathbb{R}_+} |v_{,x_3}|^2 dx_3 + c\varepsilon^{-2\mu} \int_{\mathbb{R}_+} |v|^2 dx_3. \tag{10.30}$$

To prove (10.30) we consider

$$\int_{\mathbb{R}_+} |v|^2 x_3^{-2\mu}dx_3 = \int_0^{r_0} |v|^2 x_3^{-2\mu}dx_3 + \int_{r_0}^{\infty} |v|^2 x_3^{-2\mu}dx_3 \equiv I_1 + I_2,$$

where

$$I_1 \leq c \int_0^{r_0} |v_{,x_3}|^2 x_3^{2-2\mu}dx_3 \leq cr_0^{2-2\mu} \int_0^{r_0} |v_{,x_3}|^2 dx_3$$

and

$$I_2 \leq r_0^{-2\mu} \int_{r_0}^{\infty} |v|^2 dx_3.$$

Hence

$$\int_{\mathbb{R}_+} |v|^2 x_3^{-2\mu}dx_3 \leq cr_0^{2-2\mu} \int_{\mathbb{R}_+} |v_{,x_3}|^2 dx_3 + cr_0^{-2\mu} \int_{\mathbb{R}_+} |v|^2 dx_3,$$

so (10.30) is proved. Setting $\varepsilon = |\xi|^{-1}$ and $v = \hat{u}$ we get

$$\xi^{2-2\mu} \int_{\mathbb{R}_+} |\hat{u}|^2 x_3^{-2\mu}dx_3 \leq c \int_{\mathbb{R}_+} (|\hat{u}_{,x_3}|^2 + \xi^2|\hat{u}|^2)dx_3. \tag{10.31}$$

Since $\hat{u}$ vanishes outside $(0, a)$ we can write (10.31) in the form

$$\xi^{2-2\mu} \int_0^a |\hat{u}|^2 x_3^{-2\mu}dx_3 \leq c \int_0^a (|\hat{u}_{,x_3}|^2 + \xi^2|\hat{u}|^2)dx_3. \tag{10.32}$$

From (10.29) we have

$$\xi^{2-2\mu} \int_0^a (|\hat{u}_{,x_3}|^2 + \xi^2|\hat{u}|^2)dx_3$$

$$\leq \xi^{2-2\mu} \left(\int_0^a |\hat{f}|^2 x_3^{2\mu} dx_3 \right)^{1/2} \left(\int_0^a |\hat{u}|^2 x_3^{-2\mu} dx_3 \right)^{1/2}. \tag{10.33}$$

Employing (10.32) in (10.33) yields

$$\xi^{2-2\mu} \int_0^a (|\hat{u}_{,x_3}|^2 + \xi^2|\hat{u}|^2)dx_3 \leq c \int_0^a |\hat{f}|^2 x_3^{2\mu} dx_3. \tag{10.34}$$

Hence (10.34) yields estimate for I,

$$I \leq ca_2^{2-2\mu} \int_{\mathbb{R}^2} \int_0^a |\hat{f}|^2 x_3^{2\mu} dx_3 d\xi. \tag{10.35}$$

Therefore, using the bounds for I_1, I_2, and I we obtain (10.28). This concludes the proof. □

Lemma 10.6 *Assume that $f \in L_{2,\mu}(\mathbb{R}_+^3)$, $u(0) \in H^2(\mathbb{R}^2)$, $u(0) = u|_{x_3=0}$, and $\mu \in (0,1)$. Then there exists a solution to (10.21) such that $\bar{u} = F^{-1}(\tilde{u} - \tilde{u}(0)\zeta(x_3))$, where F is the Fourier transform defined by (10.19), belongs to $H_\mu^2(\mathbb{R}_+^3)$ and the estimate holds*

$$\|\bar{u}\|_{H_\mu^2(\mathbb{R}_+^3)} \leq c(\|f\|_{L_{2,\mu}(\mathbb{R}_+^3)} + \|u(0)\|_{H^2(\mathbb{R}^2)}). \tag{10.36}$$

Proof Using (10.28) in the r.h.s. of (10.23) and assuming that a_1 is sufficiently small and a_2 is sufficiently large we derive

$$\int_{\mathbb{R}^2} \xi^2 d\xi \int_0^a (|\hat{u}_{,x_3}|^2 + \xi^2|\hat{u}|^2)dx_3 + \int_{\mathbb{R}^2} \xi^2 d\xi \int_0^a |\hat{u}|^2 x_3^{2\mu-2} dx_3$$

$$+ \int_{\mathbb{R}^2} d\xi \int_0^a (|\hat{u}_{,x_3x_3}|^2 x_3^{2\mu} + |\hat{u}_{,x_3}|^2 x_3^{2\mu-2} + |\hat{u}|^2 x_3^{2\mu-4})dx_3 \tag{10.37}$$

$$\leq c \int_{\mathbb{R}^2} d\xi \int_0^a |\hat{f}|^2 x_3^{2\mu} dx_3.$$

Employing the form of $\hat{f}$ from (10.26)$_1$ and applying the Parseval identity we derive (10.36). This concludes the proof. □

Finally, we have to estimate the last norm on the r.h.s. of (10.36). Let us recall that f defined in (10.2) is introduced in (3.4), (3.6). Hence we consider the problem (10.17) formulated in the form

$$- \Delta u = \alpha_{,x_3} \qquad x_3 > 0,$$
$$u_{,x_3} = 0 \qquad x_3 = 0,$$

(10.38)

where $\alpha = \tilde{\theta}\eta$ and $\tilde{\theta}$ is obtained from $\tilde{d}$ by the procedure formulated in (10.14)–(10.16). We assume that $\operatorname{supp}\tilde{\theta}$ is compact. Moreover,

$$\tilde{\theta}|_{x_3=0} = \theta,$$

and θ is an extension of d in a similar way as $\tilde{\theta}$ is derived from $\tilde{d}$.

Since d is given, we have the estimates

$$\|\theta\|_{H^s(\mathbb{R}^2)} \leq c\|d\|_{H^s(S_2(0))},$$
$$\|\tilde{\theta}\|_{H^{s+1/2}(\mathbb{R}^3_+)} \leq c\|\theta\|_{H^s(\mathbb{R}^2)},$$

(10.39)

$$\|\tilde{d}\|_{H^{s+1/2}(\Omega)} \leq c\|d\|_{H^s(S_2(0))}.$$

Lemma 10.7 *Assume that u is a solution to (10.38), $u(0) = u|_{x_3=0}$, α is described above, and $d \in H^1(S_2(0))$. Then*

$$\|u(0)\|_{H^2(S_2(0))} \leq c\|d\|_{H^1(S_2(0))}.$$

(10.40)

Proof Using the Neumann function, any solution to (10.38) can be expressed by

$$u(x) = \int_{\mathbb{R}^3_+} \left(\frac{1}{|x-y|} + \frac{1}{|x-\bar{y}|} \right) \alpha_{,y_3} dy,$$

where $\bar{y} = (y_1, y_2, -y_3)$.

Integrating by parts we obtain

$$u(x) = \int_{\mathbb{R}^3_+} \partial_{y_3} \left[\left(\frac{1}{|x-y|} + \frac{1}{|x-\bar{y}|} \right) \alpha \right] dy - \int_{\mathbb{R}^3_+} \partial_{y_3} \left(\frac{1}{|x-y|} + \frac{1}{|x-\bar{y}|} \right) \alpha dy$$

$$= \int_{\mathbb{R}^2} \left(\frac{1}{|x-y|} + \frac{1}{|x-\bar{y}|} \right) \Big|_{y_3=0} \tilde{\theta} dy - \int_{\mathbb{R}^3_+} \left(\frac{x_3-y_3}{|x-y|^3} - \frac{x_3+y_3}{|x-\bar{y}|^3} \right) \tilde{\theta}\eta dy$$

$$= \int_{\mathbb{R}^2} \frac{2}{\sqrt{(x'-y')^2 + x_3^2}} \tilde{\theta} dy' - \int_{\mathbb{R}^3_+} \left(\frac{x_3-y_3}{|x-y|^3} - \frac{x_3+y_3}{|x-\bar{y}|^3} \right) \tilde{\theta}\eta dy.$$

In view of this formula we have

$$\int_{\mathbb{R}^2} |\partial_{x'}^2 u(0)|^2 dx' \le 4 \int_{\mathbb{R}^2} dx' \left| \partial_{x'}^2 \int_{\mathbb{R}^2} \frac{1}{|x'-y'|} \tilde{\theta} dy' \right|^2$$

$$+4 \int_{\mathbb{R}^2} dx' \left| \partial_{x'}^2 \int_{\mathbb{R}^3_+} \frac{y_3}{\sqrt{|x'-y'|^2 + y_3^2}} \tilde{\theta} dy' dy_3 \right|^2 \le c(\|\theta\|_{H^1(\mathbb{R}^2)}$$

$$+\|\tilde{\theta}\|_{H^1(\mathbb{R}^3_+)}) \le c\|\theta\|_{H^1(\mathbb{R}^2)},$$

so in view of (10.39) we derive (10.40). This concludes the proof.   □

Chapter 11
The Neumann Problem (3.6) in L_p-Weighted Spaces

Abstract In this chapter we show W_p^2-weighted regularity, with $p \geq 3$, for solutions to the Neumann problem for the Poisson equation (3.6), i.e.

$$\Delta \varphi = -\operatorname{div} b \quad \text{in} \quad \Omega,$$

$$\bar{n} \cdot \nabla \varphi = 0 \quad \text{on} \quad S,$$

$$\int_\Omega \varphi dx = 0.$$

We also consider auxiliary problem that follows from (3.6) by reflection with respect to S_1 and localization to a neighborhood of S_2. We use the H^2-weighted regularity considered in Chap. 10. The main tool in this part is the classical technique of increasing regularity through the Marcinkiewicz-Mikhlin type result (formulated in Proposition 11.9), connected with the partition of unity which generates the weight. By the weight we mean the power of the distance to the boundary S_2 (bottom or top of the cylinder). The local estimates are possible to attain by extending the solutions on $\mathbb{R}_+^3$.

In Chap. 10 the local regularity of weak solutions to problem (3.6) is proved in weighted spaces $H_\mu^2(D)$, where $D = \{x \in \mathbb{R}_+^3 : x' \in \mathbb{R}^2, 0 < x_3 < a\}$, $\mu \in (0,1)$. The result is formulated in Lemma 10.6 by Formula (10.36). We restricted our considerations to neighborhood of $S_2(0)$ because near S_2 only the weighted spaces must be used. We replaced $S_2(-a)$ from Chap. 3 by $S_2(0)$ for convenience. In L_2-weighted spaces the proofs from Chap. 10 depend heavily on the Fourier transform (10.19). Any element $u \in H_\mu^2$, $\mu \in (0,1)$ is such that $u|_{x_3=0} = 0$, $u_{,x_3}|_{x_3=0} = 0$ but on S_2 we have either inflow or outflow. Therefore, we introduce the function $\hat{u}$ defined below (10.20). Finally, the trace $u|_{x_3=0}$ is estimated in (10.40).

© Springer Nature Switzerland AG 2019

J. Rencławowicz, W. M. Zajączkowski, *The Large Flux Problem to the Navier-Stokes Equations*, Advances in Mathematical Fluid Mechanics, https://doi.org/10.1007/978-3-030-32330-1_11

In this chapter we consider the problem

$$
\begin{aligned}
-\Delta u &= f & &\text{in } D, \\
u_{,x_3} &= 0 & &x_3 = 0, \\
u &= 0 & &x_3 = a, \\
u|_{x_3=0} &= 0,
\end{aligned}
\tag{11.1}
$$

where $u = F^{-1}\hat{u}$, $f = F^{-1}\hat{f}$, and F is defined by (10.19).

Problem (11.1) follows from the problem (3.6) by reflection with respect to S_1 and localization to a neighborhood of S_2. The localization is made by using a smooth cut-off function of x_3 equal 1 near S_2 and vanishing near $x_3 = a$. Therefore it is sometimes convenient to consider problem (11.1) for $x_3 \in \mathbb{R}_+$. Then we reckon that the solution to (11.1) is extended by 0 with respect to x_3. In fact this is a natural interpretation because this possibility follows from the properties of the cut-off function (in x_3). The role of the problem (11.1) is then to help to detect such behavior of solutions to (3.6) that the energy type estimate for solutions to problem (1.1) proved in Lemma 3.2 is attainable. Therefore integrals on $(0, a)$ and $\mathbb{R}_+$ are equivalent.

We have to mention that the methods and tools used in this chapter had already appeared in [Z10] in a different setting.

In Lemmas 10.5 and 10.6 we proved local regularity of weak solutions such that $u \in H_\mu^2(D)$ and

$$
\|u\|_{H_\mu^2(D)} \le c(\|f\|_{L_{2,\mu}(D)} + \|d\|_{H^1(S_2)}),
\tag{11.2}
$$

where $\mu \in (0, 1)$. The aim of this chapter is to show that $f \in L_{p,\mu}(D)$ implies that $u \in V_{p,\mu}^2(D)$, for $p > 2$ and $V_{p,\beta}^l(Q)$, $Q \subset \mathbb{R}_+^3$, is a set of functions with the finite norm

$$
\|u\|_{V_{p,\beta}^l(Q)} = \left(\sum_{|\alpha| \le l} \int_Q |D_x^\alpha u|^p x_3^{p(\beta+|\alpha|-l)} \, dx' dx_3 \right)^{1/p},
$$

where $x' = (x_1, x_2)$, $p \in [1, \infty]$, $\beta \in \mathbb{R}$, $\alpha = (\alpha_1, \alpha_2, \alpha_3)$ is a multiindex. We observe that

$$
V_{p,\beta}^0(Q) = L_{p,\beta}(Q), \quad V_{2,\beta}^l(Q) = H_\beta^l(Q).
$$

We recall the following local regularity result in Sobolev spaces W_p^2 (see C.B. Morrey book [M]).

Lemma 11.1 (Local Regularity) *Let*

$$B_1 = \{x' \in \mathbb{R}^2 : |x'| < r\}, \quad B_2 = \{x' \in \mathbb{R}^2 : |x'| < 2r\},$$

$$\zeta_i = \zeta_i(x_3), \ \zeta_i \in C_0^\infty(\mathbb{R}_+), \ i = 1, 2, \ \zeta_1\zeta_2 = \zeta_1,$$

$$\operatorname{supp} \zeta_2 \subset \{x_3 : \ x_3 < c_1\},$$

and ζ_1, ζ_2 *are equal* 1 *near* $x_3 = 0$. *Then for a function* $u \in W_p^2(B_2 \times \mathbb{R}_+)$ *the following inequality holds*

$$\|\zeta_1 u\|_{W_p^2(B_1 \times \mathbb{R}_+)} \le c(\|\zeta_2 \Delta u\|_{L_p(B_2 \times \mathbb{R}_+)} + \|\zeta_2 u\|_{L_{p_1}(B_2 \times \mathbb{R}_+)}), \qquad (11.3)$$

where $p_1 \in [1, \infty]$.

Definition 11.2 (Partition of Unity) Consider the families $\{\zeta_j\}_{j=-\infty}^\infty$, $\{\sigma_j\}_{j=-\infty}^\infty$, where $\zeta_j, \sigma_j \in C^\infty(\mathbb{R}_+)$ satisfy

$$\operatorname{supp} \zeta_j \subset \{x_3 : \ 2^{j-1} < x_3 < 2^{j+1}\},$$

$$\operatorname{supp} \sigma_j \subset \{x_3 : \ 2^{j-2} < x_3 < 2^{j+2}\},$$

$$\zeta_j \sigma_j = \zeta_j, \quad |\partial_{x_3}^\alpha \zeta_j| + |\partial_{x_3}^\alpha \sigma_j| \le c_\alpha 2^{-j\alpha}, \ \alpha \in \mathbb{N}_0.$$

Properties of these partitions make it possible to show the following statement.

Lemma 11.3 *Let* $\beta \in \mathbb{R}$. *Then for any function*

$$u \in W_p^2(\mathbb{R}^2 \times \{2^{j-2} < x_3 < 2^{j+2}\}),$$

the following inequality holds

$$\|\zeta_j u\|_{V_{p,\beta}^2(\mathbb{R}^2 \times \mathbb{R}_+)} \le c\|\sigma_j \Delta u\|_{L_{p,\beta}(\mathbb{R}^2 \times \mathbb{R}_+)} + c\|\sigma_j u\|_{L_{p,\beta-2}(\mathbb{R}^2 \times \mathbb{R}_+)}. \qquad (11.4)$$

Proof We define

$$B = \{x' : \ |x'| < 2\},$$

$$B_\mu = \{x' : \ |x'| < 2^{1+\mu}\},$$

$$K = \{x_3 : \ 1 < x_3 < 2\},$$

$$K_\mu = \{x_3 : \ 2^\mu < x_3 < 2^{\mu+1}\}.$$

Applying Lemma 11.1 with $p_1 = p$ we obtain

$$\sum_{|\alpha|=0}^{2} \|D^\alpha(\zeta_j u)\|_{L_p(B \times K)} \le c\|\sigma_j \Delta u\|_{L_p(B_1 \times 2K)} + c\|\sigma_j u\|_{L_p(B_1 \times 2K)},$$

where $2K = \left\{x_3 : \frac{1}{2} < x_3 < 4\right\}$. In view of scaling $x \to 2^\mu x$ we have

$$\sum_{|\alpha|=0}^{2} 2^{\mu(|\alpha|-2)} \|D^\alpha(\zeta_j u)\|_{L_p(B_\mu \times K_\mu)}$$

$$\le c\|\sigma_j \Delta u\|_{L_p(B_{\mu+1} \times 2K_\mu)} + c2^{-2\mu}\|\sigma_j u\|_{L_p(B_{\mu+1} \times 2K_\mu)},$$

where $2K_\mu = \{x_3 : 2^{\mu-1} < x_3 < 2^{\mu+2}\}$.

Now we multiply this formula by $2^{\beta\mu}$ and then raise the resulting inequality to the power p. Next, we note that $\varrho = x_3 \sim 2^\mu$ and we sum over μ to obtain

$$\sum_{|\alpha|=0}^{2} \|\varrho^{\beta+|\alpha|-2} D^\alpha(\zeta_j u)\|_{L_p(\mathbb{R}^3_+)} \le c\|\varrho^\beta \sigma_j \Delta u\|_{L_p(\mathbb{R}^3_+)}$$

$$+ c\|\varrho^{\beta-2}\sigma_j u\|_{L_p(\mathbb{R}^3_+)}.$$

Applying the definition of spaces $V_{p,\beta}^2$ and $L_{p,\beta}$ to this estimate yields (11.4). This concludes the proof.   □

Corollary 11.4 *For $\beta \in \mathbb{R}$ and u as in Lemma 11.3,*

$$\|u\|_{V_{p,\beta}^2(\mathbb{R}^3_+)} \le c\|\Delta u\|_{L_{p,\beta}(\mathbb{R}^3_+)} + c\|u\|_{L_{p,\beta-2}(\mathbb{R}^3_+)}. \tag{11.5}$$

Proof We sum up inequalities (11.4) with respect to j to obtain (11.5). This ends the proof.   □

Let

$$P(\partial_{x'}, \partial_{x_3}) = -\Delta, \quad P(i\xi, \partial_{x_3}) = -\partial_{x_3}^2 + \xi^2.$$

Let $A(\xi)$ denote the operator of the problem

$$P(i\xi, \partial_{x_3})\hat{u} = \hat{f},$$
$$\hat{u}_{,x_3}|_{x_3=0} = 0, \quad \hat{u}|_{x_3=a} = 0. \tag{11.6}$$

Lemma 11.5 *Ker $A(\xi) = 0$.*

Proof Take $\xi \ne 0$. Then every solution of homogenous Eq. (11.6)$_1$ has the form

$$\hat{u} = \alpha \sin h(|\xi|x_3) + \beta \cos h(|\xi|x_3),$$

where α, β are arbitrary parameters and $(11.6)_2$ implies the equations for α, β

$$[|\xi|\alpha \cos h(|\xi|x_3) + |\xi|\beta \sin h(|\xi|x_3)]|_{x_3=0} = 0,$$
$$[\alpha \sin h(|\xi|x_3) + \beta \cos h(|\xi|x_3)]|_{x_3=a} = 0,$$

so $\alpha = 0$, $\beta = 0$.

If $\xi = 0$, then any solution to homogeneous $(11.6)_1$ has the form

$$\hat{u} = \alpha x_3 + \beta,$$

and now $(11.6)_2$ gives $\alpha = 0$, $\beta = 0$. This concludes the proof.   □

Corollary 11.6 *There exists an inverse operator $A^{-1}(\xi)$ to problem (11.6) such that*

$$\hat{u}(\xi, x_3) = A^{-1}(\xi)\hat{f}(\xi, x_3).$$

Corollary 11.7 *From Lemma 10.6 we have for $\beta \in (0,1)$*

$$\|\hat{u}\|_{V^2_{2,\beta}(\mathbb{R}_+)} = \|A(\xi)^{-1}\hat{f}\|_{V^2_{2,\beta}(\mathbb{R}_+)} \le c\|\hat{f}\|_{V^0_{2,\beta}(\mathbb{R}_+)}. \tag{11.7}$$

Let $\hat{f}_\nu = \hat{f}\zeta_\nu$ and let $\hat{u}_\nu$ be a solution to the problem

$$P(\xi, \partial_{x_3})\hat{u}_\nu = \hat{f}_\nu,$$
$$\hat{u}_{\nu,x_3}|_{x_3=0} = 0, \quad \hat{u}_\nu|_{x_3=a} = 0. \tag{11.8}$$

Then

$$\hat{u}_\nu = A^{-1}(\xi)\hat{f}_\nu. \tag{11.9}$$

Hence

$$\sigma_\mu \hat{u}_\mu = \sigma_\mu A^{-1}(\xi)\hat{f}_\nu.$$

The further presentation depends heavily on the considerations from Maz'ya and Plamenevskii [MP, Section 7].

Lemma 11.8 *In view of Corollary 11.7 and $\xi \neq 0$ we have*

$$\|\sigma_\mu A^{-1}(\xi)\hat{f}_\nu\|_{V^0_{2,\beta}(\mathbb{R}_+)\to V^0_{2,\beta}(\mathbb{R}_+)} \le c2^{-\varepsilon|\mu-\nu|+2\mu}, \tag{11.10}$$

where $\beta \in (0,1)$ and $\varepsilon > 0$.

Proof In view of (11.7) and properties of the partition of unity, we have

$$\|\sigma_\mu \hat{u}_\nu\|_{V^2_{2,\beta}(\mathbb{R}_+)} \leq 2^{\varepsilon\mu}\|\sigma_\mu \hat{u}_\nu\|_{V^2_{2,\beta-\varepsilon}(\mathbb{R}_+)} \leq 2^{\varepsilon\mu}\|\hat{f}_\nu\|_{V^0_{2,\beta-\varepsilon}(\mathbb{R}_+)}$$

$$\leq 2^{\varepsilon(\mu-\nu)}\|\hat{f}_\nu\|_{V^0_{2,\beta}(\mathbb{R}_+)}.$$

Since

$$\|\sigma_\mu \hat{u}_\nu\|_{V^2_{2,\beta}(\mathbb{R}_+)} \geq 2^{-2\mu}\|\sigma_\mu \hat{u}_\nu\|_{V^0_{2,\beta}(\mathbb{R}_+)},$$

we conclude

$$\|\sigma_\mu \hat{u}_\nu\|_{V^2_{2,\beta}(\mathbb{R}_+)} \leq 2^{\varepsilon(\mu-\nu)+2\mu}\|\hat{f}_\nu\|_{V^0_{2,\beta}(\mathbb{R}_+)}$$

$$= 2^{-\varepsilon(\nu-\mu)+2\mu}\|\hat{f}_\nu\|_{V^0_{2,\beta}(\mathbb{R}_+)}. \tag{11.11}$$

Taking $-\varepsilon$ instead of ε we get similarly

$$\|\sigma_\mu \hat{u}_\nu\|_{V^0_{2,\beta}(\mathbb{R}_+)} \leq 2^{-\varepsilon(\mu-\nu)+2\mu}\|\hat{f}_\nu\|_{V^0_{2,\beta}(\mathbb{R}_+)}. \tag{11.12}$$

In view of (11.9), (11.11), and (11.12) we conclude the proof.   □

Now, we recall the following Marcinkiewicz-Mikhlin type result (see Dunford and Schwartz [DS, Part 2, Ch. 11, Theorem 28]).

Proposition 11.9 (Marcinkiewicz-Mikhlin Theorem) *Let $L_p(\mathbb{R}^d; H)$ be the space of functions with the finite norm*

$$\|f\|_{L_p(\mathbb{R}^d;H)} = \left(\int_{\mathbb{R}^d} \|f(z,\cdot)\|^p_H dz\right)^{1/p} < \infty,$$

where H is a Hilbert space. Let $M(\xi)$, $\xi \in \mathbb{R}^d$, be a bounded linear operator in H. Assume that for $s = 0, \cdots, d$, $i_k \neq i_l$,

$$|\xi|^s \left\|\frac{\partial^s M}{\partial \xi_{i_1} \ldots \partial \xi_{i_s}}(\xi)\right\|_{H \to H} \leq \text{const}.$$

Then, if F is the Fourier transform in $\mathbb{R}^d$, $F^{-1}_{\xi \to z} M(\xi) F_{z \to \xi}$ is a continuous operator in $L_p(\mathbb{R}^d; H)$.

Lemma 11.10 *In view of Corollary 11.7 and $u_\nu \in V^2_{2,\beta}(\mathbb{R}^3_+)$ such that*

$$P(\partial_{x'}, \partial_{x_3})u_\nu = f_\nu,$$

$$u_{\nu,x_3}|_{x_3} = 0, \quad u_\nu|_{x_3=a} = 0, \tag{11.13}$$

we have

$$\int_{\mathbb{R}^2} \left(\int_{\mathbb{R}_+} x_3^{2\beta} |\sigma_\mu(x_3) u_\nu(x', x_3)|^2 dx_3 \right)^{p/2} dx'$$

$$\leq c 2^{-p\varepsilon|\mu-\nu|+2\mu p} \int_{\mathbb{R}^2} \left(\int_{\mathbb{R}_+} x_3^{2\beta} |\zeta_\nu(x_3) f(x', x_3)|^2 dx_3 \right)^{p/2} dx'. \tag{11.14}$$

Proof We have

$$u_\nu = F^{-1}_{\xi \to x'} A(\xi)^{-1} F_{x' \to \xi} \zeta_\nu f,$$

where F denotes the Fourier transform in $\mathbb{R}^2$.

On the other hand, applying the Marcinkiewicz-Mikhlin result formulated in Proposition 11.9 we find that $F^{-1} M(\xi) F$, where $M = \sigma_\mu A(\xi)^{-1} \zeta_\nu$ is a continuous operator in $L_p(\mathbb{R}^2; V_{2,\beta}^0(\mathbb{R}_+))$. Thus using estimate (11.10) we derive the result. This concludes the proof. □

Lemma 11.11 *Let the assumptions of Corollary 11.7 hold and $u_\nu \in V_{2,\beta}^2(\mathbb{R}_+^3)$. Then for $p \geq 2$ and some $\varepsilon_1 > 0$ we have*

$$\int_{\mathbb{R}_+^3} x_3^{p(\beta-1)-2} |\zeta_\mu u_\nu(x', x_3)|^p dx' dx_3$$

$$\leq c 2^{-|\mu-\mu|\varepsilon_1 p} \int_{\mathbb{R}_+^3} x_3^{p(\beta+1)-2} |\zeta_\nu f(x', x_3)|^p dx' dx_3. \tag{11.15}$$

Proof By the Hölder inequality we estimate the integral on the r.h.s. of (11.14) as follows

$$\int_{\mathbb{R}^2} \left(\int_{\mathbb{R}_+} x_3^{2\beta} |\zeta_\nu(x_3) f(x', x_3)|^2 dx_3 \right)^{p/2} dx'$$

$$\leq \int_{\mathbb{R}^2} \left[\left(\int_{\text{supp}\,\zeta_\nu} 1^{\frac{p}{p-2}} dx_3 \right)^{\frac{p-2}{p}} \left(\int_{\mathbb{R}_+} x_3^{p\beta} |\zeta_\nu(x_3) f(x', x_3)|^p dx_3 \right)^{2/p} \right]^{p/2} dx'$$

$$\leq 2^{(\nu-1)\frac{p-2}{2}} \int_{\mathbb{R}^2} \int_{\mathbb{R}_+} x_3^{p\beta} |\zeta_\nu(x_3) f(x', x_3)|^p dx_3 dx'$$

$$\leq c \int_{\mathbb{R}^2} \int_{\mathbb{R}_+} x_3^{p(\beta+1/2)-1} |\zeta_\nu(x_3) f(x', x_3)|^2 dx_3 dx',$$

where we used that $\text{supp}\,\zeta_\nu \subset (2^{\nu-1}, 2^{\nu+1})$, $x_3 \in (2^{\nu-1}, 2^{\nu+1})$, $2^{(\nu-1)(\frac{p}{2}-1)} \sim x_3^{\frac{p}{2}-1}$ so $x_3^{\beta p + \frac{p}{2} - 1} = x_3^{p(\beta+1/2)-1}$.

Next, we deal with the l.h.s. of (11.14). Let $\{Q_j\}$ cover $\mathbb{R}^2$, where Q_j is a box with the edge equal to $2^{\mu-1}$ and by $2Q_j$ we denote a box with parallel sides to Q_j and with edges equal to 2^μ. The boxes $2Q_j$ must have the same centers as Q_j. Hence $Q_j \subset 2Q_j$. By the Hölder inequality we have

$$I_1 \equiv \int_{2Q_j} \int_{\mathbb{R}_+} |\sigma_\mu u_\nu| dx' dx_3$$

$$\leq \int_{2Q_j} \left(\int_{\mathrm{supp}\,\sigma_\mu} 1^2 dx_3 \right)^{1/2} \left(\int_{\mathbb{R}_+} |\sigma_\mu u_\nu|^2 dx_3 \right)^{1/2} dx'$$

$$\leq \int_{2Q_j} 2^{\frac{\mu}{2}+1} \left(\int_{\mathbb{R}_+} |\sigma_\mu u_\nu|^2 dx_3 \right)^{1/2} dx' \equiv I_2,$$

where we used that $\mathrm{supp}\,\sigma_\mu \subset (2^{\mu-2}, 2^{\mu+2})$. Next, we obtain

$$I_2 \leq \int_{2Q_j} 2^{\mu/2-\beta\mu} \left(\int_{\mathbb{R}_+} x_3^{2\beta} |\sigma_\mu u_\nu|^2 dx_3 \right)^{1/2} dx',$$

where we used that $x_3 \sim 2^\mu$ on $\mathrm{supp}\,\sigma_\mu$, so $2^{-2\beta\mu} x_3^{2\beta} \sim 1$ on $\mathrm{supp}\,\sigma_\mu$. Using estimates for I_1 and I_2, we have

$$I_1^p = \left(\int_{2Q_j} \int_{\mathbb{R}_+} |\sigma_\mu u_\nu| dx' dx_3 \right)^p$$

$$\leq c 2^{(1/2-\beta)p\mu} \left[\int_{2Q_j} \left(\int_{\mathbb{R}_+} x_3^{2\beta} |\sigma_\mu u_\nu|^2 dx_3 \right)^{1/2} dx' \right]^p \equiv I_3. \qquad (11.16)$$

Using the Hölder inequality we have

$$I_3 \leq c 2^{(1/2-\beta)p\mu} \left[\left(\int_{2Q_j} 1^{\frac{p}{p-1}} dx_3' \right)^{\frac{p-1}{p}} \cdot \right.$$

$$\left. \cdot \left(\int_{2Q_j} \left(\int_{\mathbb{R}_+} x_3^{2\beta} |\sigma_\mu u_\nu|^2 dx_3 \right)^{p/2} dx' \right)^{1/p} \right]^p \equiv I_4. \qquad (11.17)$$

Using that

$$\mathrm{meas}\, 2Q_j \leq c 2^{2\mu}$$

we get

$$I_4 \leq c 2^{(1/2-\beta)\mu p + 2\mu(p-1)} \int_{2Q_j} \left(\int_{\mathbb{R}_+} x_3^{2\beta} |\sigma_\mu u_\nu|^2 dx_3 \right)^{p/2} dx'. \qquad (11.18)$$

Take $|\mu - \nu| > 3$, then $\mathrm{supp}\,\zeta_\nu \cap \mathrm{supp}\,\sigma_\mu = \phi$. Then using (11.4) for $\zeta_j = \zeta_\nu$, $\sigma_j = \sigma_\mu$, $u = u_\nu$, the first term on the r.h.s. vanishes. Hence using scaling $x \to 2^\mu x$ we get

$$
\int_{Q_j} \int_{\mathbb{R}_+} |\sigma_\mu u_\nu|^p dx' dx_3 \le c 2^{3\mu(1-p)} \left(\int_{2Q_j} \int_{\mathbb{R}_+} |\sigma_\mu u_\nu| dx' dx_3 \right)^p
\tag{11.19}
$$
$$
= c 2^{3\mu(1-p)} I_1^p,
$$

because scaling on the l.h.s. gives factor $2^{3\mu}$ and on the r.h.s. it is $2^{3\mu p}$. From (11.16) to (11.18) we have

$$
I_1^p \le c 2^{(1/2 - \beta)\mu p + 2\mu(p-1)} \int_{2Q_j} \left(\int_{\mathbb{R}_+} x_3^{2\beta} |\sigma_\mu u_\nu|^2 dx_3 \right)^{p/2} dx'.
\tag{11.20}
$$

Hence, (11.19) and (11.20) imply

$$
\int_{\mathbb{R}_+^3} |\sigma_\mu u_\nu|^p dx' dx_3
$$
$$
\le c 2^{(1/2-\beta)\mu p + 2\mu(p-1) + 3\mu(1-p)} \int_{\mathbb{R}^2} \left(\int_{\mathbb{R}_+} x_3^{2\beta} |\sigma_\mu u_\nu|^2 dx_3 \right)^{p/2} dx',
\tag{11.21}
$$

where

$$
\left(\frac{1}{2} - \beta \right) \mu p + 2\mu(p-1) + 3\mu(1-p) = -\mu \left[p \left(\beta + \frac{1}{2} \right) - 1 \right].
$$

Next, inequalities (11.14) and (11.21) give

$$
2^{\mu[p(\beta - 3/2) - 1]} \int_{\mathbb{R}_+^3} |\sigma_\mu u_\nu|^p dx' dx_3
$$
$$
\le 2^{\mu[p(\beta - 3/2) - 1]} 2^{-\mu[p(\beta + 1/2) - 1]} \cdot 2^{-p\varepsilon|\mu - \nu| + 2\mu p}.
$$
$$
\cdot \int_{\mathbb{R}^2} \left(\int_{\mathbb{R}_+} x_3^{2\beta} |\zeta_\nu(x_3) f(x', x_3)|^2 dx_3 \right)^{p/2} dx'
\tag{11.22}
$$
$$
\le c 2^{-p\varepsilon|\mu - \nu|} \int_{\mathbb{R}^2} \left(\int_{\mathbb{R}_+} x_3^{2\beta} |\zeta_\nu(x_3) f(x', x_3)|^2 dx_3 \right)^{p/2} dx'
$$
$$
\le c 2^{-p\varepsilon|\mu - \nu|} \int_{\mathbb{R}_+^3} x_3^{p(\beta + 1/2) - 1} |\zeta_\nu f|^p dx_3 dx',
$$

where we used the Hölder inequality

$$\int_{\mathbb{R}^2}\left(\int_{\mathbb{R}_+} x_3^{2\beta}|\zeta_\nu f|^2 dx_3\right)^{p/2} dx'$$

$$\leq \int_{\mathbb{R}^2}\left[\left(\int_{\operatorname{supp}\zeta_\nu} 1 dx_3\right)^{\frac{p-2}{p}}\left(\int_{\mathbb{R}_+} x_3^{p\beta}|\zeta_\nu f|^p dx_3\right)^{2/p}\right]^{p/2} dx'$$

$$\leq c\int_{\mathbb{R}^2}\int_{\mathbb{R}_+\cap\operatorname{supp}\zeta_\nu} 2^{\nu\left(\frac{p}{2}-1\right)} x_3^{p\beta}|\zeta_\nu f|^p dx_3$$

$$\leq \int_{\mathbb{R}_+^3} x_3^{p(\beta+1/2)-1}|\zeta_\nu f|^p dx_3 dx'.$$

We multiply both sides of (11.22) by $2^{\mu(p/2-1)}$, use that $\operatorname{supp}\zeta_\mu \subset \operatorname{supp}\sigma_\mu$, and on the l.h.s. we exploit properties of $\operatorname{supp}\zeta_\mu$ but on the r.h.s. properties of $\operatorname{supp}\zeta_\nu$. Therefore, (11.22) yields

$$2^{\mu(p/2-1)}2^{\mu[p(\beta-3/2)-1]}\int_{\mathbb{R}_+^3}|\zeta_\mu u_\nu|^p dx' dx_3$$

$$\leq c2^{\mu(p/2-1)}2^{-p\varepsilon|\mu-\nu|}\int_{\mathbb{R}_+^3} x_3^{p(\beta+1/2)-1}|\zeta_\nu f|^p dx' dx_3.$$

Continuing, we have

$$2^{\mu[p(\beta-1)-2]}\int_{\mathbb{R}_+^3}|\zeta_\mu u_\nu|^p dx' dx_3$$

$$\leq c2^{(\mu-\nu)(p/2-1)-p\varepsilon|\mu-\nu|}\int_{\mathbb{R}_+^3} x_3^{p(\beta+1)-2}|\zeta_\nu f|^p dx' dx_3, \tag{11.23}$$

where $x_3^{-(p/2-1)}|_{\operatorname{supp}\zeta_\nu} = c2^{-\nu(p/2-1)}$.

Let $\mu > \nu$. Then the exponent of 2 on the r.h.s. equals

$$-p(\mu-\nu)\left[\varepsilon-\frac{1}{2}+\frac{1}{p}\right] \equiv -p(\mu-\nu)\varepsilon_1,$$

and for $\mu < \nu$ it equals

$$-p\varepsilon(\nu-\mu)\left[\varepsilon+\frac{1}{2}-\frac{1}{p}\right] = -p(\nu-\mu)\varepsilon_1.$$

Hence (11.23) takes the form

$$
\int_{\mathbb{R}^3_+} x_3^{p(\beta-1)-2}|\zeta_\mu u_\nu|^p dx' dx_3
$$

$$
\leq c2^{-p|\mu-\nu|\varepsilon_1} \int_{\mathbb{R}^3_+} x_3^{p(\beta+1)-2}|\zeta_\nu f|^p dx' dx_3.
$$

(11.24)

We note that in the case $|\mu - \nu| < 3$ we have to add on the r.h.s. of (11.19) the expression

$$
c2^{2\mu p} \int_{2Q_j} \int_{\mathbb{R}_+} |\zeta_\nu f|^p dx' dx_3.
$$

(11.25)

Multiply (11.25) by the same factor as in (11.22) we obtain (11.25) in the form

$$
c2^{2\mu p}2^{\mu[p(\beta-3/2)-1]}2^{-\nu[p(\beta+1)-2]} \int_{\mathbb{R}^3_+} x_3^{p(\beta+1)-2}|\zeta_\nu f|^p dx' dx_3,
$$

where the factor before the integral equals

$$
c2^{(\mu-\nu)[p(\beta+1/2)-1]}2^{-\nu(p/2-1)} \leq c
$$

because $p > 2$ and $|\mu - \nu| < 3$.

The above considerations imply (11.15) and this concludes the proof. □

Estimate (11.15) has a local character. To obtain this estimate for whole $\mathbb{R}^3_+$ we need the following result from [MP, Sect. 7, Lemma 7.7]. Let $\{\zeta_j\}_{j=-\infty}^{\infty}$ be a partition of unity introduced in Definition 11.2. Let $\mathcal{E}_0$ and $\mathcal{E}_1$ be two Banach spaces such that the multiplication of their elements by smooth scalar functions is defined. Assume that there exist numbers p and q, $1 \leq p \leq q \leq \infty$, such that for all $u \in \mathcal{E}_0$ and $v \in \mathcal{E}_1$ the following inequalities are valid

$$
\|u\|_{\mathcal{E}_0} \leq c\left(\sum_{j=-\infty}^{\infty} \|\zeta_j u\|_{\mathcal{E}_0}^q \right)^{1/q}
$$

(11.26)

and

$$
\|v\|_{\mathcal{E}_1} \geq c\left(\sum_{j=-\infty}^{\infty} \|\zeta_j v\|_{\mathcal{E}_1}^p \right)^{1/p},
$$

(11.27)

where $\| \; \|_{\mathcal{E}_i}$ is the norm of $\mathcal{E}_i$, $i = 0, 1$.

Lemma 11.12 (See [MP, Section 7, Lemma 7.7]) *Let $\vartheta : \mathcal{E}_1 \to \mathcal{E}_0$ be a linear operator defined on functions with compact supports such that for some $\varepsilon > 0$ and arbitrary integers μ and ν we have*

$$\|\zeta_\mu \vartheta \zeta_\nu v\|_{\mathcal{E}_0} \le c e^{-\varepsilon|\mu-\nu|}\|\zeta_\nu v\|_{\mathcal{E}_1}, \qquad (11.28)$$

where v is an arbitrary element of $\mathcal{E}_1$. Then for any $v \in \mathcal{E}_1$ with a compact support the following inequality holds

$$\|\vartheta v\|_{\mathcal{E}_0} \le c\|v\|_{\mathcal{E}_1}, \qquad (11.29)$$

where c does not depend on v.

Proof In view of (11.26),

$$\|\vartheta v\|_{\mathcal{E}_0} = \left\|\vartheta\left(\sum_{\nu=-\infty}^{\infty} \zeta_\nu v\right)\right\|_{\mathcal{E}_0} \le c\left(\sum_{\mu=-\infty}^{\infty}\left\|\sum_{\nu=-\infty}^{\infty} \zeta_\mu \vartheta \zeta_\nu v\right\|_{\mathcal{E}_0}^q\right)^{1/q}$$
$$\le c\left[\sum_{\mu=-\infty}^{\infty}\left(\sum_{\nu=-\infty}^{\infty} \|\zeta_\mu \vartheta \zeta_\nu v\|_{\mathcal{E}_0}\right)^q\right]^{1/q}.$$

Hence, from (11.28) we derive

$$\|\vartheta v\|_{\mathcal{E}_0} \le c\left[\sum_{\mu=-\infty}^{\infty}\left(\sum_{\nu=-\infty}^{\infty} e^{-\varepsilon|\mu-\nu|}\|\zeta_\nu v\|_{\mathcal{E}_1}\right)^q\right]^{1/q}.$$

Employing the properties of the discrete convolution we obtain for $q \ge p$

$$\|\vartheta v\|_{\mathcal{E}_0} \le c\left(\sum_{\nu=-\infty}^{\infty} \|\zeta_\nu v\|_{\mathcal{E}_1}^p\right)^{1/p}.$$

Hence (11.27) implies (11.29) and concludes the proof.   □

Lemma 11.13 *Let u be a solution to the problem*

$$P(\partial_{x'}, \partial_{x_3})u = f,$$
$$u_{,x_3}|_{x_3=0} = 0, \quad u|_{x_3=a} = 0. \qquad (11.30)$$

Let $f \in L_{\beta+1-2/p}(\mathbb{R}^3_+)$, $p \ge 2$. Then

$$\int_{\mathbb{R}^3_+} x_3^{p(\beta-1)-2}|u(x', x_3)|^p dx' dx_3 \le c\int_{\mathbb{R}^3_+} x_3^{p(\beta+1)-2}|f(x', x_3)|^p dx' dx_3.$$
$$(11.31)$$

Proof To prove the lemma we apply Lemma 11.12 to inequalities (11.15). Setting $q = p$, $\mathcal{E}_0(\mathbb{R}^3_+) = V^0_{p,\beta-1-2/p}(\mathbb{R}^3_+)$, $\mathcal{E}_1(\mathbb{R}^3_+) = V^0_{p,\beta+1-2/p}(\mathbb{R}^3_+)$, and assuming that ϑ is the inverse operator to the operator of problem (11.30), we write (11.15) in the form

$$\|\zeta_\mu \vartheta \zeta_\nu v\|_{\mathcal{E}_0(\mathbb{R}^3_+)} \le c \exp(-\varepsilon|\mu - \nu|)\|\zeta_\nu v\|_{\mathcal{E}_1(\mathbb{R}^3_+)}.$$

Then Lemma 11.12 implies (11.31). This concludes the proof. □

Lemma 11.14 *Let u be a solution to (11.30). Let $f \in V^0_{p,\varkappa}(\mathbb{R}^3_+)$, $p \ge 2$. Then*

$$\|u\|_{V^2_{p,\varkappa}(\mathbb{R}^3_+)} \le c\|f\|_{V^0_{p,\varkappa}(\mathbb{R}^3_+)}. \tag{11.32}$$

Proof In view of the regularity result (11.5) we have

$$\|u\|_{V^2_{p,\varkappa}(\mathbb{R}^3_+)} \le c\|f\|_{V^0_{p,\varkappa}(\mathbb{R}^3_+)} + c\|u\|_{V^0_{p,\varkappa-2}(\mathbb{R}^3_+)}. \tag{11.33}$$

Now, we apply (11.31). Since $f \in V^0_{p,\varkappa}(\mathbb{R}^3_+)$ we have that $p(\beta + 1) - 2 = p\varkappa$ so $\beta = \varkappa - 1 + 2/p$. Then $p(\beta - 1) - 2 = p(\varkappa - 2)$. Therefore, (11.31) yields

$$\|u\|_{V^0_{p,\varkappa-2}(\mathbb{R}^3_+)} \le c\|f\|_{V^0_{p,\varkappa}(\mathbb{R}^3_+)}.$$

Combining this with (11.33) implies (11.32). This concludes the proof. □

Chapter 12
Existence of Solutions (v, p) and (h, q)

Abstract In the last chapter we use the estimates proved in the previous parts of the book to show the existence of solutions (v, p) to Navier-Stokes equations in cylindrical domains. In Sect. 12.1 we use the Galerkin approximation to establish the existence of weak solutions such that v is weakly continuous with respect to t in $L^2(\Omega)$ norm and v converges to $v_0 = v|_{t=0}$ as $t \to 0$ strongly in $L^2(\Omega)$ norm, as well as global weak solutions such that

$$v \in V(\Omega \times (kT, (k+1)T)), \quad \forall k \in \mathbb{N}_0 = \mathbb{N} \cup \{0\}, \quad \text{where}$$

$$\|v\|_{V(\Omega \times (kT,(k+1)T))} = \operatorname*{ess\,sup}_{kT<t<(k+1)T} |v(t)|_{L_2(\Omega)} + |\nabla v|_{L_2(\Omega \times (kT,(k+1)T))}.$$

To prove the large time existence of regular solutions in Sect. 12.2 we construct the corresponding mapping to apply Leray-Schauder fixed point theorem. We show this for such solutions that $v \in W_2^{2,1}(\Omega^T)$, $\nabla p \in L_2(\Omega^T)$, and with $h = v_{,x_3}, q = p_{,x_3}$,

$$\|h\|_{W_\sigma^{2,1}(\Omega^t)} + \|\nabla q\|_{L_\sigma(\Omega^t)} \leq cD_9, \quad 5/3 \leq \sigma \leq 10/3,$$

for some constant D_9 which depends on data and $T > 0$.

In this chapter, we sketch the proofs of existence results.

12.1 Existence of Weak Solutions

We prove the existence of weak solutions to problem (1.1) with Galerkin method.

© Springer Nature Switzerland AG 2019
J. Renclawowicz, W. M. Zajączkowski, *The Large Flux Problem to the Navier-Stokes Equations*, Advances in Mathematical Fluid Mechanics, https://doi.org/10.1007/978-3-030-32330-1_12

Theorem 12.1 *Assume the compatibility condition (1.4). Let*

$$v(0) \in L_2(\Omega), f \in L_2(0, T; L_{6/5}(\Omega)),$$

$$d_i \in L_\infty(0, T; W_p^{s-1/p}(S_2)) \cap L_2(0, T; W_2^{1/2}(S_2)),$$

$$\text{where } \frac{3}{p} + \frac{1}{3} \le s, p > 3 \text{ or } p = 3, s > \frac{4}{3},$$

$$d_{i,t} \in L_2(0, T; W_{6/5}^{1/6}(S_2)), \ i = 1, 2.$$

Then there exists a weak solution v to problem (1.1) such that v is weakly continuous with respect to t in $L^2(\Omega)$ norm and v converges to v_0 as $t \to 0$ strongly in $L^2(\Omega)$ norm. Moreover, $v \in V(\Omega^T)$, $v \cdot \bar{\tau}_\alpha \in L_2(0, T; L_2(S_1))$, for $\alpha = 1, 2$ and v satisfies, for all $t \le T$

$$\|v\|_{V(\Omega^t)}^2 + \gamma \sum_{\alpha=1}^{2} \int_0^t \|v \cdot \bar{\tau}_\alpha\|_{L_2(S_1)}^2 \le 2\|f\|_{L_2(0,t;L_{6/5}(\Omega))}^2$$

$$+\varphi \left(\sup_{\tau \le t} \|d\|_{W_3^{s-1/p}(S_2)} \right) \left(\|d\|_{L_2(0,t;W_2^{1/2}(S_2))}^2 + \|d_t\|_{L_2(0,t;W_{6/5}^{1/6}(S_2))}^2 \right) \tag{12.1}$$

$$+\|v(0)\|_{L_2(\Omega)}^2,$$

where φ is a nonlinear positive increasing function.

With the a priori estimate, we can show also the existence of global weak solutions.

Theorem 12.2 *Assume the compatibility condition (1.4). Let*

$$f \in L_2(kT, (k+1)T; L_{6/5}(\Omega)),$$

$$d_i \in L_\infty(\mathbb{R}^+; W_p^{s-1/p}(S_2)) \cap L_2(kT, (k+1)T; W_2^{1/2}(S_2)),$$

$$\text{where } \frac{3}{p} + \frac{1}{3} \le s, p > 3 \text{ or } p = 3, s > \frac{4}{3},$$

$$d_{i,t} \in L_2(kT, (k+1)T; W_{6/5}^{1/6}(S_2)), i = 1, 2.$$

Let us assume that

$$\|v(0)\|_{L_2(\Omega)} \le A$$

for some constant A and

$$2\int_{kT}^{(k+1)T} \|f\|^2_{L_{6/5}(\Omega)} + \varphi\left(\sup_t \|d\|_{W_p^{s-1/p}(S_2)}\right)$$

$$\cdot \int_{kT}^{(k+1)T}\left(\|d\|^2_{W_2^{1/2}(S_2)} + \|d_t\|^2_{W_{6/5}^{1/6}(S_2)}\right) \le (1 - e^{-\nu T})A^2$$

for all $k \in \mathbb{N}_0$, where φ is a nonlinear positive increasing function. Then there exists a global weak solution v to (1.1) such that

$$v \in V(\Omega \times (kT, (k+1)T)) \quad \forall k \in \mathbb{N}_0 = \mathbb{N} \cup \{0\},$$

and

$$\|v\|^2_{V(\Omega \times (kT,t))} \le 2\int_{kT}^t \|f\|^2_{L_{6/5}(\Omega)}d\tau + A^2 \quad (12.2)$$

$$+\varphi\left(\sup_\tau \|d\|_{W_p^{s-1/p}(S_2)}\right)\int_{kT}^t\left(\|d\|^2_{W_2^{1/2}(S_2)} + \|d_t\|^2_{W_{6/5}^{1/6}(S_2)}\right)d\tau$$

for $t \in (kT, (k+1)T]$.

We use the estimate from Lemma 3.2

$$\|w\|^2_{V(\Omega \times (kT,t))} \le c\mathcal{A}^2(T) + \exp(-\nu kT)|w(0)|^2_{2,\Omega}, \quad (12.3)$$

where

$$\mathcal{A}^2(T) = \sup_{k \in \mathbb{N}_0} \int_{kT}^{(k+1)T} (|f(t)|^2_{6/5,\Omega} + \|d(t)\|^2_{1,3,S_2}$$

$$+\|d_t(t)\|^2_{6/5,S_2})dt\, \varphi(\sup_t \|d(t)\|_{1,3,S_2}) < \infty,$$

and (w, p) is a solution to the problem (3.8):

$$w_t + w \cdot \nabla w + w \cdot \nabla \delta + \delta \cdot \nabla w - \operatorname{div} \mathbb{T}(w, p)$$

$$= f - \delta_t - \delta \cdot \nabla \delta + \nu \operatorname{div} \mathbb{D}(\delta) \equiv F(\delta, t) \quad \text{in } \Omega^T,$$

$$\operatorname{div} w = 0 \quad \text{in } \Omega^T,$$

$$w \cdot \bar{n} = 0 \quad \text{on } S^T,$$

$$\nu\bar{n} \cdot \mathbb{D}(w) \cdot \bar{\tau}_\alpha + \gamma w \cdot \bar{\tau}_\alpha = -\nu\bar{n} \cdot \mathbb{D}(\delta) \cdot \bar{\tau}_\alpha - \gamma\delta \cdot \bar{\tau}_\alpha$$

$$\equiv B_{1\alpha}(\delta), \quad \alpha = 1, 2, \quad \text{on } S_1^T,$$

$$\bar{n} \cdot \mathbb{D}(w) \cdot \bar{\tau}_\alpha = -\bar{n} \cdot \mathbb{D}(\delta) \cdot \bar{\tau}_\alpha \equiv B_{2\alpha}(\delta), \quad \alpha = 1, 2, \quad \text{on } S_2^T,$$

$$w|_{t=0} = v(0) - \delta(0) \equiv w(0) \quad \text{in } \Omega.$$

To prove the existence of weak solutions to the problem (3.8) we use the Galerkin method. We follow the ideas of Ladyzhenskaya [L1, Chapter 6, Sect.7]. Namely, we introduce a sequence of approximating functions w^N given as

$$w^N(x, t) = \sum_{k=1}^{N} C_{kN}(t) a^k(x),$$

where $\{a^k\}_{k=1}^{\infty}$ is a system of orthonormal functions in $L^2(\Omega) \cap J_2^0(\Omega)$. Here,

$$J_2^0(\Omega) = \{f \in H^1(\Omega) : \operatorname{div} f = 0\}$$

and $\{a^k\}_{k=1}^{\infty}$ is a fundamental system in $H^1(\Omega)$ satisfying

$$\sup_{k \in \mathbb{N}} \sup_{x \in \Omega} |a^k(x)| < \infty, \quad \sup_{k \in \mathbb{N}} \sup_{x \in \partial\Omega} |a^k(x)| < \infty.$$

The coefficients $C_{kN}(0)$ are defined by

$$C_{kN}|_{t=0} = (w_0, a_k), \quad k = 1, \ldots, N,$$

and the functions w^N satisfy the following system with test functions a^k:

$$\left\{ \int_{\Omega} \left(\frac{1}{2} \frac{d}{dt} w^N a^k + w^N \cdot \nabla w^N a^k + \delta \cdot \nabla w^N \cdot a^k + w^N \cdot \nabla \delta \cdot a^k \right. \right.$$

$$\left. + \nu \mathbb{D}(w^N) \mathbb{D}(a^k) \right) dx + \gamma \int_{S_1} w^N \cdot \bar{\tau}_j a^k \bar{\tau}_j dS_1 \bigg\}$$

$$= \left(\sum_{j,\sigma=1}^{2} \int_{S_\sigma} B_{\sigma j} a^k \cdot \bar{\tau}_j \, dS_\sigma + \int_{\Omega} F \cdot a^k dx \right)$$

for $k = 1, \ldots, N$. Thus, w^N are weak solutions to (3.8).

With

$$(f, g) = \int_{\Omega} f g \, dx$$

and

$$(f, g)_S = \int_{S} f g \, dS$$

this can be rewritten as:

$$\left\{ (w_t^N, a^k) + (w^N \cdot \nabla w^N, a^k) + (\delta \cdot \nabla w^N, a^k) + (w^N \cdot \nabla \delta, a^k) \right.$$

$$\left. + \nu(\mathbb{D}(w^N), \mathbb{D}(a^k)) + \gamma(w^N \cdot \bar{\tau}_j, a^k \cdot \bar{\tau}_j)_{S_1} \right\} =$$

$$\left[\sum_{\sigma,j=1}^{2} (B_{\sigma j}, a^k \cdot \bar{\tau}_j)_{S_\sigma} + (F, a^k) \right], \quad k = 1, \dots, N.$$

Thus,

$$\left(\frac{d}{dt} w^N, a^k \right) + (w^N \cdot \nabla w^N, a^k) + (\delta \cdot \nabla w^N, a^k) + (w^N \cdot \nabla \delta, a^k)$$

$$+ \nu(\mathbb{D}(w^N), \mathbb{D}(a^k)) + \gamma(w^N \cdot \bar{\tau}_j, a^k \cdot \bar{\tau}_j)_{S_1} \quad (12.4)$$

$$= \sum_{j,\sigma=1}^{2} (B_{\sigma j}, a^k \cdot \bar{\tau}_j)_{S_\sigma} + (F, a^k), \quad k = 1, \dots, N.$$

The above equations are in fact a system of ordinary differential equations for the functions $C_{kN}(t)$. The properties of the sequence a^k imply

$$\|w^N(x,t)\|_{L_2(\Omega)}^2 = \sum_{k=1}^{N} C_{kN}^2(t).$$

On the other hand, we can obtain a priori bounds for the approximate solutions w^N of the same form as (12.3):

$$\|w^N\|_{V(\Omega^T)}^2 = \sup_{0 \le t \le T} \|w^N\|_{L_2(\Omega)} + \int_0^T \|\nabla w^N\|_{L_2(\Omega)} dt$$

$$\le \varphi \left(\sup_{0 \le t \le T} \|\tilde{d}\|_{W_p^s(\Omega)} \right) \int_0^T \left(\|\tilde{d}\|_{W_2^1(\Omega)}^2 + \|\tilde{d}_t\|_{W_{6/5}^1(\Omega)}^2 \right) dt \quad (12.5)$$

$$\int_0^T \|f\|_{L_{6/5}(\Omega)}^2 dt + \|w^N(0)\|_{L_2(\Omega)}^2 \le C,$$

where $\frac{3}{p} + \frac{1}{3} \le s, p > 3$ or $p = 3, s > \frac{4}{3}$. Therefore, $\sup_{0 \le t \le T} |C_{kN}(t)|$ is bounded on $[0, T]$ and w^N are well defined for all times t.

Let us now define $\psi_{N,k} \equiv (w^N(x,t), a^k(x))$. This sequence is uniformly bounded by (12.5). We can also show that it is equicontinuous. Namely, we integrate (12.4) with respect to t from t to $t + \Delta t$ to obtain

$$|\psi_{N,k}(t+\Delta t)-\psi_{N,k}(t)| \leq \sup_{x\in\Omega}|a^k(x)|\int_t^{t+\Delta t}\left(|w^N\cdot\nabla w^N|_{L_2(\Omega)}+|\delta\cdot\nabla w^N|_{L_2(\Omega)}\right.$$

$$+|w^N\cdot\nabla\delta|_{L_2(\Omega)}+|F|_{L_2(\Omega)}\Big)\,d\tau+\nu|\nabla a^k|_{L_2(\Omega)}\int_t^{t+\Delta t}|\nabla w^N|_{L_2(\Omega)}d\tau$$

$$+\gamma\sup_{x\in S}|a^k(x)|\int_t^{t+\Delta t}\left(|w^N\cdot\bar{\tau}_j|_{L_2(S_1)}+\sum_{j,\sigma=1}^{2}|B_{\sigma j}|_{L_2(S_\sigma)}\right)d\tau$$

$$\leq \sup_{x\in\Omega}|a^k(x)|\sqrt{\Delta t}\left(\sup_{x\in\Omega}|w^N|_{L_2(\Omega)}(|\nabla w^N|_{L_2(\Omega^T)}+|\nabla\delta|_{L_2(\Omega^T)})\right.$$

$$+\sup_{x\in\Omega}|\delta|_{L_2(\Omega)}|\nabla w^N|_{L_2(\Omega^T)}\Bigg)$$

$$+\sup_{x\in\Omega}|a^k(x)|\int_t^{t+\Delta t}|F|_{L_2(\Omega)}d\tau+\nu|\nabla a^k|_{L_2(\Omega)}\sqrt{\Delta t}|\nabla w^N|_{L_2(\Omega^T)}$$

$$+\gamma\sup_{x\in S}|a^k(x)|\left(\sqrt{\Delta t}|\nabla w^N|_{L_2(\Omega^T)}+\int_t^{t+\Delta t}\sum_{j=1}^{2}|B_j|_{L_2(S)})d\tau\right)$$

$$\leq C(k)\left(\sqrt{\Delta t}+\int_t^{t+\Delta t}(|F|_{L_2(\Omega)}+\sum_{j=1}^{2}|B_j|_{L_2(S)})d\tau\right).$$

We can see that for given k and $N\geq k$ the r.h.s. tends to zero as $\Delta t\to 0$ uniformly in N. Thus, it is possible to choose a subsequence N_m such that $\psi_{N_m,k}$ converges as $m\to\infty$ uniformly to some continuous function ψ_k for any given k. Since the limit function w is defined as

$$w(x,t)=\sum_{k=1}^{\infty}\psi_k(t)a^k(x),$$

we conclude that $(w^{N_m}-w,\psi(x))$ tends to zero as $m\to\infty$ uniformly with respect to $t\in[0,T]$ for any $\psi\in J_2^0(\Omega)$ and $w(x,t)$ is continuous in t in weak topology. Moreover, estimates (12.5) remain true for the limit function w.

We will show that $\{w^{N_m}\}$ converges strongly in $L_2(\Omega^T)$. To this end, we need to apply the following version of the Friedrichs lemma: for any $\varepsilon>0$, there exists N_ε such that for any $u\in W_2^1(\Omega)$ (see Lemma 6.1 from Ladyzhenskaya et al. [LSU, Ch. 5, Sect. 6]) the following inequality holds:

$$\|u\|_{L_2(\Omega)}^2\leq\sum_{k=1}^{N_\varepsilon}|(u,a^k)|^2+\varepsilon\|\nabla u\|_{L_2(\Omega)}^2.$$

This in terms of $u = w^{N_m} - w^{N_l}$ reads

$$\|w^{N_m} - w^{N_l}\|^2_{L_2(\Omega^T)} \le \sum_{k=1}^{N_\varepsilon} \int_0^T |(w^{N_m} - w^{N_l}, a^k)|^2 dt + \varepsilon \|\nabla w^{N_m} - \nabla w^{N_l}\|^2_{L_2(\Omega^T)}.$$

By (12.5), we have

$$\|\nabla w^{N_m} - \nabla w^{N_l}\|^2_{L_2(\Omega^T)} \le 2C^2$$

for some constant C. The first integral on the r.h.s. for a given number N_ε can be arbitrarily small if only m and l are sufficiently large, so it tends to zero as $m, l \to \infty$. Therefore, $\{w^{N_m}\}$ converges strongly in $L_2(\Omega^T)$.

We summarize the above convergence properties of the sequence $\{w^{N_m}\}$:

1. $\{w^{N_m}\} \to w$ strongly in $L_2(\Omega^T)$ for some w,
2. $\{w^{N_m}\} \to w$ weakly in $L_2(\Omega)$ uniformly with respect to $t \in [0, T]$,
3. $\nabla\{w^{N_m}\} \to \nabla w$ weakly in $L_2(\Omega^T)$.

For given $\Phi^k = \sum_{j=1}^k d_j(t) a^j(x)$, the sequence $\{w^{N_m}\}$ satisfies the identities

$$\int_\Omega \left(\frac{d}{dt} w^{N_m} \Phi^k + (w^{N_m} \cdot \nabla w^{N_m} + \delta \cdot \nabla w^{N_m} + w^{N_m} \cdot \nabla \delta) \Phi^k + \nu \mathbb{D}(w^{N_m}) \mathbb{D}(\Phi^k) \right) dx$$

$$+ \gamma \int_{S_1} w^{N_m} \cdot \bar\tau_j \Phi^k \cdot \bar\tau_j dS_0 = \sum_{\sigma,j=1}^2 \int_{S_\sigma} B_{\sigma j} \Phi^k \cdot \bar\tau_j dS_\sigma + \int_\Omega F \Phi^k dx.$$

Then, we can pass to the limit with $m \to \infty$ to obtain the identity for w. The conditions $\operatorname{div} w^N = 0, w^N \cdot \bar n|_{S_T} = 0$ remain true for the limit function w as well.

It remains to consider the limit $\lim_{t \to 0} w(x, t)$. We note that w^{N_m} satisfies the relation (3.10) (if we use the test function w^{N_m}). This yields

$$\|w^{N_m}\|_{L_2(\Omega)} \le \|w_0\|_{L_2(\Omega)} + \int_0^t (\|F\|_{L_2(\Omega)} + \|B\|_{L_2(S)}) dt.$$

In the limit $m \to \infty$ we obtain

$$\|w\|_{L_2(\Omega)} \le \|w_0\|_{L_2(\Omega)} + \int_0^t (\|F\|_{L_2(\Omega)} + \|B\|_{L_2(S)}) dt,$$

which implies

$$\overline{\lim}_{t \to 0} \|w\|_{L_2(\Omega)} \le \|w_0\|_{L_2(\Omega)}.$$

On the other hand, since w^{N_m} tends to w as $m \to \infty$, we have $\|w^{N_m} - w_0\|_{L_2(\Omega)} \to 0$. Therefore, $|w^{N_m} - w_0| \to 0$ weakly in $L^2(\Omega)$ as $t \to 0$ and

$$\|w_0\|_{L_2(\Omega)} \leq \underline{\lim}_{t \to 0} \|w\|_{L_2(\Omega)}.$$

We conclude that the limit $\lim_{t \to 0} \|w\|_{L_2(\Omega)}$ exists and is equal to $\|w_0\|_{L_2(\Omega)}$ where the convergence is strong, in the $L_2(\Omega)$ norm.

Consequently, we have proved the following result.

Lemma 12.3 *Let the assumptions of Theorem 12.1 be satisfied. Then there exists a weak solution w to problem (3.8) such that w is weakly continuous with respect to t in the $L_2(\Omega)$ norm and w converges to w_0 as $t \to 0$ strongly in the $L_2(\Omega)$ norm.*

Since $v = w - \delta$ we deduce the analogous existence result for v, formulated in Theorem 12.1.

12.2 Existence of Regular Solutions

Lemma 12.4 *Let the assumptions of Theorem 4.14 be satisfied. Then the solution (v, p) to (1.1) and solution (h, q) to (4.6) satisfying (4.81) and (4.82) exist on $(0, T)$ for some $T > 0$.*

Proof To prove the existence of solutions to problem (1.1) we will use the Leray-Schauder theorem. To this end, we construct the mappings

$$
\begin{aligned}
v_t - \operatorname{div} \mathbb{T}(v, p) = -\lambda \tilde{v} \cdot \nabla \tilde{v} + f & \quad \text{in } \Omega^T = \Omega \times (0, T), \\
\operatorname{div} v = 0 & \quad \text{in } \Omega^T, \\
v \cdot \bar{n} = 0 & \quad \text{on } S_1^T, \\
\nu \bar{n} \cdot \mathbb{D}(v) \cdot \bar{\tau}_\alpha + \gamma v \cdot \bar{\tau}_\alpha = 0, \ \alpha = 1, 2, & \quad \text{on } S_1^T, \\
v \cdot \bar{n} = d & \quad \text{on } S_2^T, \\
\bar{n} \cdot \mathbb{D}(v) \cdot \bar{\tau}_\alpha = 0, \ \alpha = 1, 2, & \quad \text{on } S_2^T, \\
v\big|_{t=0} = v(0) & \quad \text{in } \Omega,
\end{aligned}
\tag{12.6}
$$

and

$$h_{,t} - \operatorname{div} \mathbb{T}(h, q) = -\lambda(\tilde{v} \cdot \nabla \tilde{h} + \tilde{h} \cdot \nabla \tilde{v}) + g \quad \text{in } \Omega^T,$$

$$\operatorname{div} h = 0 \quad \text{in } \Omega^T,$$

$$n \cdot \bar{h} = 0, \quad \bar{n} \cdot \mathbb{D}(h) \cdot \bar{\tau}_\alpha + \gamma h \cdot \bar{\tau}_\alpha = 0, \quad \alpha = 1, 2 \quad \text{on } S_1^T, \tag{12.7}$$

$$h_i = -d_{x_i}, \quad i = 1, 2, \quad h_{3,x_3} = \Delta' d \quad \text{on } S_2^T,$$

$$h\big|_{t=0} = h(0) \quad \text{in } \Omega,$$

where $g = f_{,x_3}, \Delta' = \partial_{x_1}^2 + \partial_{x_2}^2$, $\lambda \in [0,1]$ and $\tilde{v}$, $\tilde{h}$ are treated as given functions. We assume that $\tilde{h} = \tilde{v}_{,x_3}$, thus differentiating (12.6) with respect to x_3 and subtracting from (12.7) we obtain that

$$h = v_{,x_3}.$$

Problems (12.6) and (12.7) define the mappings

$$\Phi_1 : (\tilde{v}, \lambda) \to (v, p),$$

$$\Phi_2 : (\tilde{v}, \tilde{h}, \lambda) \to (h, q).$$

We set $\Phi = (\Phi_1, \Phi_2)$. In the previous chapters we have shown a priori estimate for a fixed point of Φ for $\lambda = 1$. On the other hand, for $\lambda = 0$ we have a unique existence of solutions to problems (12.6) and (12.7). Let us introduce the space

$$\mathcal{M}(\Omega^T) = L_{2r}(0, (T; W^2_{\frac{6\eta}{3+\eta}}(\Omega)), \quad \eta \geq 2, \quad r \geq 2.$$

We shall find restrictions on r, η such that

$$\Phi : \mathcal{M}(\Omega^{(T)}) \times \mathcal{M}(\Omega^T) \to \mathcal{M}(\Omega^T) \times \mathcal{M}(\Omega^T)$$

is a compact mapping.

Assume that $\tilde{v} \in L_{2r}(0, T; W^1_{\frac{6\eta}{3+\eta}}(\Omega))$. Then

$$\|\tilde{v} \cdot \nabla \tilde{v}\|_{L_r(0,T;L_\eta(\Omega))} = \left(\int_0^T dt \|\tilde{v} \cdot \nabla \tilde{v}\|^r_{L_\eta(\Omega)} \right)^{1/r}$$

$$\leq \left(\int_0^T dt \|\tilde{v}\|^r_{L_{\frac{6\eta}{3-\eta}}(\Omega)} \|\nabla \tilde{v}\|^r_{L_{\frac{6\eta}{3+\eta}}(\Omega)} \right)^{1/r} \tag{12.8}$$

$$\leq c \left(\int_0^T dt \|\tilde{v}\|^{2r}_{W^1_{\frac{6\eta}{3+\eta}}(\Omega)} \right)^{1/r} \leq c \|\tilde{v}\|^2_{L_{2r}(0,T;W^1_{\frac{6\eta}{3+\eta}}(\Omega))}.$$

In the same way we obtain

$$\|\tilde{v} \cdot \nabla \tilde{h}\|_{L_r(0,T;L_\eta(\Omega))} + \|\tilde{h} \cdot \nabla \tilde{v}\|_{L_r(0,T;L_\eta(\Omega))}$$
$$\leq c\|\tilde{v}\|_{L_{2r}(0,T;W^1_{\frac{6\eta}{3+\eta}}(\Omega))}\|\tilde{h}\|_{L_{2r}(0,T;W^1_{\frac{6\eta}{3+\eta}}(\Omega))}. \tag{12.9}$$

In view of (12.8) and (12.9) can be shown that solutions to problems (12.6) and (12.7) belong to $W^{2,1}_{\eta,r}(\Omega^T)$ (see Solonnikov [S1]).

We are going to use the following imbeddings

$$W^{2,1}_2(\Omega^T) \supset W^{2,1}_{\eta,r}(\Omega^T) \tag{12.10}$$

and

$$W^{2,1}_2(\Omega^T) \subset L_{2r}(0,T;W^1_{\frac{6\eta}{3+\eta}}(\Omega)) \equiv \mathcal{M}(\Omega^T), \tag{12.11}$$

where (12.10) holds for $\eta \geq 2$, $r \geq 2$ and (12.11) is compact for r,η satisfying the inequality

$$\frac{5}{2} - \frac{3}{\frac{6\eta}{3+\eta}} - \frac{2}{2r} < 1,$$

which takes the form

$$1 < \frac{3}{2\eta} + \frac{1}{r}. \tag{12.12}$$

Setting $r = \eta = 2$ we obtain that $\tilde{v}, \tilde{h} \in L_4(0,T;W^1_{12/5}(\Omega))$ and then condition (12.12) takes the form

$$1 < \frac{3}{4} + \frac{1}{2} \quad \text{so} \quad \frac{1}{2} < \frac{3}{4}. \tag{12.13}$$

Hence, we have compactness of mappings Φ_1 and Φ_2.

To show the continuity of mappings Φ_1 and Φ_2 we consider

$$v_{st} - \operatorname{div} \mathbb{T}(v_s, p_s) = -\lambda \tilde{v}_s \cdot \nabla \tilde{v}_s + f \qquad \text{in } \Omega^T,$$
$$\operatorname{div} v_s = 0 \qquad \text{in } \Omega^T,$$
$$\bar{n} \cdot v_s = 0, \quad \nu \bar{n} \cdot \mathbb{D}(v_s) \cdot \bar{\tau}_\alpha + \gamma v_s \cdot \bar{\tau}_\alpha = 0 \qquad \text{on } S_1^T,$$
$$v_s \cdot \bar{n} = d \qquad \text{on } S_2^T,$$
$$\bar{n} \cdot \mathbb{D}(v_s) \cdot \bar{\tau}_\alpha = 0, \quad \alpha = 1,2, \qquad \text{on } S_2^T,$$
$$v_s|_{t=0} = v(0) \qquad \text{in } \Omega$$

and

$$h_{st} - \operatorname{div} \mathbb{T}(h_s, q_s) = -\lambda(\tilde{h}_s \cdot \nabla \tilde{v}_s + \tilde{v}_s \cdot \nabla \tilde{h}_s) + g \quad \text{in } \Omega^T$$

$$\operatorname{div} h_s = 0 \quad \text{in } \Omega^T$$

$$\bar{n} \cdot h_s = 0, \quad \nu \bar{n} \cdot \mathbb{D}(h_s) \cdot \bar{\tau}_\alpha + \gamma h_s \cdot \bar{\tau}_\alpha = 0, \quad \alpha = 1, 2 \quad \text{on } S_1^T,$$

$$h_{si} = -d_{x_i}, \quad i = 1, 2, \quad h_{s3, x_3} = \Delta' d \quad \text{on } S_2^T,$$

$$h_s|_{t=0} = h(0) \quad \text{in } \Omega,$$

where $s = 1, 2$.
Let

$$V = v_1 - v_2, \quad H = h_1 - h_2, \quad P = p_1 - p_2, \quad Q = q_1 - q_2.$$

Then V and H are solutions to the problems

$$V_t - \operatorname{div} \mathbb{T}(V, P) = -\lambda(\tilde{V} \cdot \nabla \tilde{v}_1 + \tilde{v}_2 \cdot \nabla \tilde{V})$$

$$\operatorname{div} V = 0$$

$$V \cdot \bar{n}|_{S_1} = 0, \quad \nu \bar{n} \cdot \mathbb{D}(V) \cdot \bar{\tau}_\alpha + \gamma V \cdot \bar{\tau}_\alpha|_{S_1} = 0, \quad \alpha = 1, 2, \quad (12.14)$$

$$V \cdot \bar{n}|_{S_2} = d \ , \bar{n} \cdot \mathbb{D}(V) \cdot \bar{\tau}_\alpha|_{S_2} = 0, \quad \alpha = 1, 2,$$

$$V|_{t=0} = 0,$$

$$H_t - \operatorname{div} \mathbb{T}(H, Q) = -\lambda(\tilde{H} \cdot \nabla \tilde{v}_1 + \tilde{h}_2 \cdot \nabla \tilde{V} + \tilde{V} \cdot \nabla \tilde{h}_1 + \tilde{v}_2 \cdot \nabla \tilde{H})$$

$$\operatorname{div} H = 0$$

$$H \cdot \bar{n}|_{S_1} = 0, \quad \nu \bar{n} \cdot \mathbb{D}(H) \cdot \bar{\tau}_\alpha + \gamma H \cdot \bar{\tau}_\alpha|_{S_1} = 0, \quad \alpha = 1, 2, (12.15)$$

$$H_i|_{S_2} = -d_{x_i}, \quad i = 1, 2, \quad H_{3, x_3}|_{S_2} = 0,$$

$$H|_{t=0} = 0.$$

Assume that $\lambda \neq 0$. In Theorem 4.14, we proved the existence of such a constant $\mathcal{D}$ that for sufficiently chosen data holds

$$\|h\|_{W_2^{2,1}(\Omega^T)} \leq cD_9 \equiv \mathcal{D} \quad \text{and} \quad \|v\|_{W_2^{2,1}(\Omega^T)} \leq \varphi(D_2, D_9, \mathcal{A}) + cD_8' \equiv c\mathcal{D}.$$

Then for solutions of (12.14) we have

$$\|V\|_{\mathcal{M}(\Omega^T)} = \|V\|_{L_{2r}(0, T; W_{\frac{6\eta}{3+\eta}}^1(\Omega))} \leq c\|V\|_{W_2^{2,1}(\Omega^T)} \leq c\|V\|_{W_{\eta, r}^{2,1}(\Omega^T)}$$

$$(12.16)$$

$$\leq c \sum_{s=1}^{2} \|\tilde{v}_s\|_{L_{2r}(0; T; W_{\frac{6\eta}{3+\eta}}^1(\Omega))} \cdot \|\tilde{V}\|_{L_{2r}(0; T; W_{\frac{6\eta}{3+\eta}}^1(\Omega))} \leq c(\mathcal{D})\|\tilde{V}\|_{\mathcal{M}(\Omega^T)},$$

where $r \geq 2$, $\eta \geq 2$, and satisfy either (12.12) or (12.13).

Similarly we can prove that

$$\|H\|_{\mathcal{M}(\Omega^T)} \leq c(\mathcal{D})(\|\tilde{V}\|_{\mathcal{M}(\Omega^T)} + \|\tilde{H}\|_{\mathcal{M}(\Omega^T)}). \qquad (12.17)$$

Inequalities (12.16) and (12.17) imply continuity of mapping Φ. Continuity with respect to λ is obvious. Hence by the Leray-Schauder fixed point theorem we have the existence of solutions to problem (1.1) such that $v \in W_2^{2,1}(\Omega^T)$, $\nabla p \in L_2(\Omega^T)$ and estimates (4.81) and (4.82) hold. This concludes the proof. $\qquad\square$

References

[Ad] Adams, R.: Sobolev Spaces, vol. 65, 268 pp. Academic, New York (1975)

[ADN] Agmon, S., Douglis, A., Nirenberg, L.: Estimates near the boundary for solutions of elliptic partial differential equations satisfying general boundary conditions, part I. Commun. Pure Appl. Math. **12**, 623–727 (1959); part II. Commun. Pure Appl. Math. **17**, 25–92 (1964)

[AF] Adams, R., Fournier, J.: Sobolev Spaces, vol. 140, 2nd edn., 305 pp. Elsevier, Amsterdam (2003)

[B] Bugrov, Ya.S.: Imbedding theorems for classes of functions with mixed norms. Mat. Sb. Number 4(12) **92**(134), 611–621 (1973, in Russian)

[BIN] Besov, O.V., Il'in, V.P., Nikolskii, S.M.: Integral Representations of Functions and Imbedding Theorems, vol. I. Scripta Series in Mathematics. V.H. Winston, New York (1978)

[BMN1] Babin, A., Mahalov, A., Nicolaenko, B.: Global regularity of 3D rotating Navier-Stokes equations for resonant domains. Appl. Math. Lett. **13**, 51–57 (2000)

[BMN2] Babin, A., Mahalov, A., Nicolaenko, B.: Regularity and integrability of 3D Euler and Navier-Stokes equations for rotating fluids. Asymptot. Anal. **15**, 103–150 (1997)

[BMN3] Babin, A., Mahalov, A., Nicolaenko, B.: Global regularity of 3D rotating Navier-Stokes equations for resonant domains. Indiana Univ. Math. J. **48**, 1136–1176 (1999)

[CF] Coifman, R., Fefferman, C.: Weighted norm inequalities for maximal functions and singular integrals. Studia Math. **51**, 241–250 (1974)

[DS] Dunford, N., Schwartz, J.T.: Linear Operators. Wiley, New York (1966)

[FST] Farwig, R., Schulz, R., Taniuchi, Y.: Spatial asymptotic profiles of solutions to the Navier-Stokes system in a rotating frame with fast decaying data. Hokkaido Math. J. **47**(3), 501–529 (2018)

[FVSU] Fadeev, D.K., Vulikh, B.Z., Solonnikov, V.A., Uraltseva, N.N.: Some chapters of analysis and higher algebra, Publ. of St. Petersburg University, 200 pp. (1981)

[G] Golovkin, K.K.: On equivalent norms for fractional spaces. Trudy Mat. Inst, Steklov **66**, 364–383 (1962, in Russian); English transl.: Am. Math. Soc. Transl. (2) **81**, 257–280 (1969)

© Springer Nature Switzerland AG 2019
J. Renclawowicz, W. M. Zajączkowski, *The Large Flux Problem to the Navier-Stokes Equations*, Advances in Mathematical Fluid Mechanics, https://doi.org/10.1007/978-3-030-32330-1

[G1] Galdi, G.P.: An introduction to the mathematical theory of the Navier-Stokes equations. In: Linear Steady Problems. Vol. I: Springer Tracts in Natural Philosophy, vol. 38, 475 pp. Springer, Berlin (1994); revised edition: 1998

[G2] Galdi, G.P.: An introduction to the mathematical theory of the Navier-Stokes equations. In: Nonlinear Steady Problems. Vol. II: Springer Tracts in Natural Philosophy, vol. 39, 364 pp. Springer, New York (1994)

[IS] Il'in, V.P., Solonnikov, V.A.: On some properties of differentiable functions of many arguments. Trudy Mat. Inst, Steklov. **66**, 205–226 (1962, in Russian); English transl.: Am. Math. Soc. Transl. **81**, 67–90 (1969)

[K] Kondratiev, V.A.: Boundary value problems for elliptic equations in domains with conical and angular points. Trudy Mosk, Mat. Obshch. **16**, 209–292 (1967, in Russian); English transl.: Trans. Moscow Math. Soc. **16**, 227–313 (1967)

[KP] Kapitanskii, L., Pileckas, K.: Some problems of vector analysis. Boundary value problems of mathematical physics and related problems in the theory of functions, 16. Zap. Nauchn. Sem. Leningrad. Otdel. Mat. Inst. Steklov. (LOMI) **138**, 65–85 (1984, Russian); English transl.: J. Sov. Math. **32**, 469–483 (1986)

[KZ] Kubica, A, Zajączkowski, W.M.: A priori estimates in weighted spaces for solutions of the Poisson and heat equations. Appl. Math. **34**(4), 431–444 (2007)

[L1] Ladyzhenskaya, O.A.: Mathematical Theory of Viscous Incompressible Flow. Nauka, Moscow (1970, in Russian); Second English edition, revised and enlarged, translated by Richard A. Silverman and John Chu. Mathematics and Its Applications, vol. 2, xviii+224 pp. Gordon and Breach, Science Publishers, New York

[L2] Ladyzhenskaya, O.A.: Solution "in the large" of the nonstationary boundary value problem for the Navier-Stokes system with two space variables. Dokl. Akad. Nauk SSSR **123**, 427–429 (1958); English transl.: Commun. Pure Appl. Math. **12**, 427–433 (1959)

[L3] Ladyzhenskaya, O.A.: Unique global solvability of the three-dimensional Cauchy problem for the Navier-Stokes equations in the presence of axial symmetry. Zap. Nauchn. Sem. LOMI **7**, 155–177 (1968, in Russian)

[Le] Lemarié-Rieusset, P.G.: Recent Developments in the Navier-Stokes Problem. Research Notes in Mathematics Series, vol. 431, 408 pp. Chapman & Hall/CRC, Boca Raton (2002)

[Le2] Lemarié-Rieusset, P.G.: The Navier-Stokes Problem in the 21st Century, xxii+718 pp. CRC Press, Boca Raton (2016)

[LSU] Ladyzhenskaya, O.A., Solonnikov, V.A., Uraltseva, N.N.: Linear and Quasilinear Equations of Parabolic Type. Nauka, Moscow (1967, in Russian); Translated from the Russian by S. Smith. Translations of Mathematical Monographs, vol. 23, xi+648 pp. American Mathematical Society, Providence (1968)

[M] Morrey, C.B.: Multiple Integrals in the Calculus of Variations. Springer, Berlin (1966)

[MN] Mahalov, A., Nicolaenko, B.: Global solvability of the three-dimensional Navier-Stokes equations with uniformly large initial vorticity. Usp. Mat. Nauk **58**, 2(350), 79–110 (2003, in Russian); English transl.: Russ. Math. Surv. **58**(2), 287–318 (2003)

[MP] Maz'ya, V.G., Plamenevskii, B.A.: L_p-estimates for solutions to elliptic boundary value problems in domains with edges. Trudy Moskov. Mat. Obshch. **37**, 49–93 (1978, in Russian); English transl.: Trans. Moscow Math. Soc. (1), 49–97 (1980)

[NZ1]   Nowakowski, B., Zajączkowski, W.M.: Very weak solutions to the boundary-value problem for the homogeneous heat equation. J. Anal. Appl. **32**(2), 129–153 (2013)

[NZ2]   Nowakowski, B., Zajączkowski, W.M.: Global existence of solutions to Navier-Stokes equations in cylindrical domains. Appl. Math. **36**(2), 169–182 (2009)

[PZ]   Pileckas, K., Zajączkowski, W.M.: Global solvability for large flux of a three-dimensional time dependent Navier-Stokes problem in a straight cylinder. Math. Methods Appl. Sci. **31**, 1607–1633 (2008)

[RRS]   Robinson, J.C., Rodrigo, J.L., Sadowski, W.: The three-dimensional Navier-Stokes equations. In: Classical Theory. Cambridge Studies in Advanced Mathematics, vol. 157, xiv+471 pp. Cambridge University Press, Cambridge (2016)

[RZ1]   Rencławowicz, J., Zajączkowski, W.M.: Existence of solutions to the Poisson equation in L_2-weighted spaces. Appl. Math. **37**(3), 309–323 (2010)

[RZ2]   Rencławowicz, J., Zajączkowski, W.M.: Existence of solutions to the Poisson equation in L_p-weighted spaces. Appl. Math. **37**(1), 1–12 (2010)

[RZ3]   Rencławowicz, J., Zajączkowski, W.M.: Existence of global weak solutions for Navier-Stokes equations with large flux. J. Differ. Equ. **251**, 688–707 (2011)

[RZ4]   Rencławowicz, J., Zajączkowski, W.M.: Nonstationary flow for the Navier-Stokes equations in a cylindrical pipe. Math. Methods Appl. Sci. **35**, 1434–1455 (2012)

[RZ5]   Rencławowicz, J., Zajączkowski, W.M.: Global regular solutions to the Navier-Stokes equations with large flux. Discret. Contin. Dyn. Syst. Suppl. **2011(special)**, 1234–1243 (2011)

[RZ6]   Rencławowicz, J., Zajączkowski, W.M.: Global nonstationary Navier-Stokes motion with large flux. SIAM J. Math. Anal. **46**(4), 2581–2613 (2014)

[RZ7]   Rencławowicz, J., Zajączkowski, W.M.: Large time regular solutions to the Navier-Stokes equations in cylindrical domains. Topol. Methods Nonlin. Anal. **32**, 69–87 (2008)

[S]   Sohr, H.: The Navier-Stokes Equations: An Elementary Functional Analytic Approach, 367 pp. Birkhäuser, Zurich (2001)

[S1]   Solonnikov, V.A.: On general boundary value problems for the Douglis-Nirenberg elliptic systems, Part 1. Izv. Akad Nauk SSSR, Ser. Mat. **28**, 665–706 (1964, in Russian); English transl.: Am. Math. Soc. Transl. (2) **56**, 193–232 (1966); Part 2. Trudy Matem. Inst. Steklov **92**, 233–297 (1966, in Russian); English transl.: Proc. Steklov Inst. Math. **92**, 269–339 (1968)

[S2]   Solonnikov, V.A.: A priori estimates for solutions of second order parabolic equations. Trudy Mat. Inst. Steklov **70**, 133–212 (1964, in Russian); English transl.: Am. Math. Soc. Transl. (2) **65**, 51–137 (1967)

[SS]   Solonnikov, V.A., Shchadilov, V.E.: On boundary value problem for a stationary system of the Navier-Stokes equations. Trudy Mat. Inst. Steklova **125**, 196–210 (1973, in Russian); English transl.: Proc. Steklov Inst. Math. **125**, 186–199 (1973)

[St]   Stein, E.M., Singular Integrals and Differentiability Properties of Functions. Princeton University Press, Princeton (1970)

[SZ]   Solonnikov, V.A., Zajączkowski, W.M.: On the Neumann problem for elliptic equations of the second order in domains with edges on the boundary. Zapiski Nauchn. Sem. LOMI **127**, 7–48 (1983, in Russian); English transl.: J. Soviet. Math.; J. Math. Sci. **27**, 2561–2586 (1984)

[SZ1]   Socała, J., Zajączkowski, W.M.: Long time existence of solutions to 2D Navier-Stokes equations with heat convection. Appl. Math. **36**(4), 453–463 (2009)

[SZ2]   Socała, J., Zajączkowski, W.M.: Long time estimate of solutions to 3D Navier-Stokes equations coupled with heat convection. Appl. Math. **39**(1), 23–41 (2012)

[SZ3]   Socała, J., Zajączkowski, W.M.: Long time existence of regular solutions to 3D Navier-Stokes equations coupled with heat convection. Appl. Math. **39**(2), 231–242 (2012)

[T]   Takeshita, A.: A remark on Leray's inequality. Pac. J. Math. **157**, 151–158 (1993)

[UY]   Ukhovskii, M.R., Yudovich V.I.: Axially symmetric motions of ideal and viscous fluids filling whole space. Prikl. Mat. Mekh. **32**, 59–69 (1968, in Russian)

[Z1]   Zajączkowski, W.M.: Global regular nonstationary flow for the Navier-Stokes equations in a cylindrical pipe. Topol. Methods Nonlinear Anal. **26**, 221–286 (2005)

[Z2]   Zajączkowski, W.M.: Global regular solutions to the Navier-Stokes equations in a cylinder. Banach Center Publ. **74**, 235–255 (2006)

[Z3]   Zajączkowski, W.M.: Global special regular solutions to the Navier-Stokes equations in a cylindrical domain without the axis of symmetry. Topol. Methods Nonlin. Anal. **24**, 69–105 (2004)

[Z4]   Zajączkowski, W.M.: Existence and regularity of solutions of some elliptic system in domains with edges. Discret. Math. **274**, 91 (1988)

[Z5]   Zajączkowski, W.M.: On global regular solutions to the Navier-Stokes equations in cylindrical domains. Topol. Methods Nonlin. Anal. **37**, 55–85 (2011)

[Z6]   Zajączkowski, W.M.: Long time existence of regular solutions to Navier-Stokes equations in cylindrical domains under boundary slip conditions. Studia Math. **169**, 243–285 (2005)

[Z7]   Zajączkowski, W.M.: Stability of two-dimensional solutions to the Navier-Stokes equations in cylindrical domains under Navier boundary conditions. J. Math. Anal. Appl. **444**(1), 275–297 (2016)

[Z8]   Zajączkowski, W.M.: Nonstationary Stokes system in cylindrical domains under boundary slip conditions. J. Math. Fluid Mech. **19**, 1–16 (2017)

[Z9]   Zajączkowski, W.M.: Existence of solutions to the (rot-div)-system in L_2 weighted spaces. Appl. Math. **36**(1), 83–106 (2009)

[Z10]   Zajączkowski, W.M.: Existence of solutions to the (rot-div)-system in L_p weighted spaces. Appl. Math. **37**(2), 127–142 (2010)

[Z11]   Zajączkowski, W.M.: Global special regular solutions to the Navier-Stokes equations in a cylindrical domains under boundary slip conditions, Gakuto international series. Math. Sci. Appl. **21**, 188 (2004)

[Z12]   Zajączkowski, W.M.: Global special regular solutions to the Navier-Stokes equations in axially symmetric domains under boundary slip conditions. Discret. Math. **432**, 138 (2005)

Notation Index

$D(v)$ - dilatation tensor, 2
$\mathbb{T}(v,p)$ - stress tensor, 2
Ω - cylindrical domain, 2
$\bar{n}$ - unit outward vector normal to boundary, 2
$S = \partial\Omega = S_1 \cup S_2$ - boundary of cylinder, 2
$L = \partial S_2$ - 16

functions

d, inflow-outflow function, 2
$w, \delta, v = w + \delta$ 3
$h = v_{,x_3}, q = p_{,x_3}, g = f_{,x_3}$,   4
$\chi(x,t) = v_{2,x_1} - v_{1,x_2}$ - vorticity, 4
$v' = (v_1, v_2)$,   4
φ - increasing positive function, 4
$\eta(\sigma, \varphi, \rho)$ - Hopf function, 31
$\tilde{h}$ - equation (4.14),   65
$k = h - \tilde{h}$,   67

$D_1(t) = |d_1|_{3,6,S_2^t}$,   4
$\mathcal{V}(t) = |\nabla v|_{3,2,\Omega^t}$,   4
$\mathcal{A} = \mathcal{A}(t),\ t \le T$,   4

$$\mathcal{A}^2(t) = \varphi(\sup_t \|d\|_{1,3,S_2}) \int_0^t (|f|_{6/5,\Omega}^2 + \|d\|_{1,3,S_2}^2 + \|d_t\|_{1,6/5,S_2}^2)dt' + |v(0)|_{2,\Omega}^2$$

$\Lambda_1(t), \Lambda_2(t) = \Lambda_1 + |h(0)|_{2,\Omega}$,   4

© Springer Nature Switzerland AG 2019
J. Rencławowicz, W. M. Zajączkowski, *The Large Flux Problem to the
Navier-Stokes Equations*, Advances in Mathematical Fluid Mechanics,
https://doi.org/10.1007/978-3-030-32330-1

$$\Lambda_1^2(T) = \int_0^T (\|d_{x'}\|_{1,S_2}^2 + \|d_t\|_{1,S_2}^2 + |f_3|_{4/3,S_2}^2 + |g|_{6/5,\Omega}^2)dt + \mathcal{A}^2 \sup_t \|d_{x'}\|_{1,3/2,S_2}^2$$

$H_2 = H_1 + |h|_{10/3,\Omega^t}$, $H_1 = \sup_t |h|_{3,\Omega} + \|h\|_{1,2,\Omega^t}$

$S_2(a_1, \varrho) = \{x \in \Omega : x_3 \in (-a, -a + \varrho)\}$

$S_2(a_2, \varrho) = \{x \in \Omega : x_3 \in (a - \varrho, a)\}$

$\tilde{S}_2'(a_1, \varrho) = \{x \in \Omega : x_3 \in (-a + r, -a + \varrho)\}$

$\tilde{S}_2'(a_2, \varrho) = \{x \in \Omega : x_3 \in (a - \varrho, a - r)\}$

$$F^2(t) = c[|f|_{6/5,\Omega}^2 + \varphi(\|\tilde{d}\|_{W_{3,\infty}^1(\Omega)})(\|\tilde{d}\|_{W_{3,\infty}^1(\Omega)}^2 + \|\tilde{d}_t\|_{1,6/5,\Omega}^2)]$$

$$F_1(kT, t) = \varphi(\sup_t \|d(t)\|_{1,3,S_2}) \cdot \|d\|_{\frac{1}{2},2,S_2 \times (kT,t)}^2$$

$$\Gamma^2(t) = c[|f|_{6/5,\Omega}^2 + |f_t|_{6/5,\Omega}^2 + \|\tilde{d}_{tt}\|_{1,6/5,\Omega}^2 + \|\tilde{d}_t\|_{1,2,\Omega}^2 + \|\tilde{d}\|_{1,2,\Omega}^4$$

$$+ \|\tilde{d}_t\|_{1,2,\Omega}^4] + c\varphi(\|\tilde{d}\|_{W_{3,\infty}^1(\Omega)}, \|\tilde{d}_t\|_{W_{3,\infty}^1(\Omega)}) \cdot [|\tilde{d}_{tt}|_{6/5,\infty,\Omega}^2$$

$$+ |\tilde{d}_t|_{2,\infty,\Omega}^2 + |\tilde{d}|_{2,\infty,\Omega}^4 + |\tilde{d}_t|_{2,\infty,\Omega}^4 + \|\tilde{d}\|_{W_{3,\infty}^1(\Omega)} + \|\tilde{d}_t\|_{1,6/5,\Omega}^2]$$

$$B_1^2(T) = \sup_{k \in \mathbb{N}_0} \exp \left(\int_{kT}^{(k+1)T} |w(t)|_{\infty,\Omega}^2 dt \right) \cdot \int_{kT}^{(k+1)T} (F^2(t) + \Gamma^2(t))dt$$

$$D_2 = (1+\mathcal{A})(|F_3|_{6/5,2,\Omega^t} + |\chi(0)|_{2,\Omega}) + |f|_{5/3,\Omega^t} + \|d\|_{W_{5/3}^{7/5,7/10}(S_2^t)} + \|v(0)\|_{W_{5/3}^{4/5}(\Omega)}$$

$D_3 = \mathcal{A}(1 + \mathcal{A})H_2 + \varphi(\mathcal{A}) + D_2$

$$M = D_4 H_2 + D_5 = c\gamma D_3,$$

$$D_4 = c\gamma(1 + \mathcal{A})\mathcal{A},$$

$$D_5 = c\gamma(\varphi(\mathcal{A}) + D_2), \quad D_6 = D_4^2 + (1 + \mathcal{A})D_4,$$

$$D_7 = D_4(\varphi(\mathcal{A}) + D_5 + D_2)D_5 + D_4 D_5 + (1 + \mathcal{A})D_5,$$

$$D_8 = (\varphi(\mathcal{A}) + D_5 + D_2)D_5 + c(|f|_{2,\Omega^t} + \|v(0)\|_{1,\Omega} + \|d\|_{W_2^{2-1/2,1-1/4}(S_2^t)}).$$

$D_9(t)$, 80

$$D_9(t) = \sum_{i=1}^{2} \|d_{,x_i}\|_{W_\sigma^{2-\frac{1}{\sigma},1-\frac{1}{2\sigma}}(S_2^t)} + \|\Delta'd\|_{W_\sigma^{1-\frac{1}{\sigma},\frac{1}{2}-\frac{1}{2\sigma}}(S_2^t)} + |g|_{\sigma,\Omega^t} + \|h(0)\|_{2-\frac{2}{\sigma},\sigma,\Omega}$$

$E_\Omega(u)$, 26

$$E_\Omega(u) = |\mathbb{D}(u)|_{2,\Omega}^2 = \int_\Omega (u_{i,x_j} + u_{j,x_i})^2 dx$$

spaces   - Chapter 2, 11–16
$L_2(G), L_2(G)$
$L_p(G)$
$L_{p,\mu}^k(\Omega)$
$L_q(0,t;W_p^k(\Omega)), L_p(0,t;W_p^k(\Omega))$
$V^s(\Omega^T), V(\Omega^T), V(\Omega^t), V(t_1,t_2)$
$W_r^{l,l/2}(\Omega^T)$
$W_r^{l,l/2}(S^T)$
$B_{r,\theta}^{l,l/2}(G)$
$B_{r,\theta}^{l,l_0}(G)$
$B_{\bar{r},\theta}^{\bar{l}}(\Omega)$
$H_r^{l,l_0}(G)$
$V_{p,\beta}^l(Q)$
$H_\mu^k(\Omega_\varrho)$
$L_{p,\mu}^k(\Omega)$

Name Index

Subject Index

equations:

Navier-Stokes equations, 1, 2, 7, 8, 9
Neumann problem, 3, 17, 33, 129
Poisson equation, 3, 17, 33, 129
Stokes system, 5, 22, 23

functions:

cut-off function, 65
delta Dirac function, 18
Green function, 17
Hopf function, 3, 32
vorticity, 4

spaces:

anisotropic Sobolev, 11, 12
Banach, 153
Besov, 11, 15
energy type, 14
Nikolskii, 15
Sobolev, 3, 11
Sobolev-Slobodetskii, 14
weighted, 12

theorems, lemmas, estimates:

Calderón-Zygmund estimates, 19
compatibility condition, 3
Dirichlet boundary condition, 3
Faedo-Galerkin method, 4, 157
Fourier transform, 129
Fredholm theorem, 21
Friedrichs lemma, 162
Green theorem, 35
Hölder inequalities, 35
inverse trace theorem, 24
interpolation inequality, 24
Korn inequality, 24
Leray-Schauder fixed point theorem, 4, 157, 164
Marcinkiewicz-Mikhlin theorem, 148
Muckenhoupt weight, 4, 36
Neumann boundary condition, 3
Parseval identity, 139
partition of unity, 16
Poincaré inequality, 22
regularizer technique, 19
trace theorem, 24
Young inequalities, 46

© Springer Nature Switzerland AG 2019
J. Rencławowicz, W. M. Zajączkowski, *The Large Flux Problem to the
Navier-Stokes Equations*, Advances in Mathematical Fluid Mechanics,
https://doi.org/10.1007/978-3-030-32330-1

Printed in the United States
By Bookmasters